Jack Campbell

From one [illegible] historian
to an [illegible]

[illegible]
Sept 1998

G ⑥ ~~t2~~
Nature

A Whaling Enterprise

BUSINESS

I'm sent the Genius of the British Isle,
Your long bewild'red Thoughts to unbeguile,
Enrich yourselves, may Britain smile in Peace,
Your proper Business is your Trades increase;
The Sea's your native strength, Fish is your Wealth,
In other Projects you amuse yourself.
Imploy your Poor in fishing up that Oar,
To fill your Checquer, and secure your Shoar.
Herrings, and Whales, the Mine is given you,
Search them they are your happier Peru.
In wooden Walls reign Monarchs of the Seas,
And stretch Commerce and Conquest as you please.

From a print, 1738

A Whaling Enterprise

Salvesen in the Antarctic

GERALD ELLIOT

MICHAEL RUSSELL

First published in Great Britain 1998
by Michael Russell (Publishing) Ltd
Wilby Hall, Wilby, Norwich NR16 2JP

Typeset in Sabon by The Typesetting Bureau
Allen House, East Borough, Wimborne, Dorset
Printed and bound in Great Britain
by Biddles Ltd, Guildford and King's Lynn

ISBN 0 85955 241 1

Contents

Foreword

The Antarctic whaling industry, which spanned the first seventy years of this century, has been well written up as a story of enterprise, romance and courage, or alternatively as one of destruction, greed and cruelty. No one could or need compete with Tønnessen and Johnsen's masterly four-volume history *Den Moderne Hvalfangsts Historie*, abridged in English as *The History of Modern Whaling*, and over the years many whalers have published their own experiences.

Whaling, however, can produce endless fascinating material. Each whaling enterprise has its own saga, not least my own company, Christian Salvesen, which was both a pioneer and leader in the industry, with Harold Salvesen emerging as its forceful genius. I have tried to leaven that record with my own memoirs from within the company. Descriptions of technological invention in an industry which has probably gone for ever are unlikely to spark enthusiasm, but the ways of life of whaling, still proudly remembered forty years on, are part of Norwegian and Scottish history. The rise and fall of the industry provide plenty of business lessons, and the international efforts to conserve the whale stocks, though disastrously unsuccessful, are as relevant today for fish and other natural resources as they were then.

One: 1907-1931

This is a chronicle of the grandeur, decline and eventual extinction of the Antarctic whaling industry. It is linked with the fortunes of the Norwegian-Scottish company Christian Salvesen, for half a century one of the leading venturers in Antarctic whaling and closely concerned in the international struggles, eventually fruitless, to preserve the Antarctic whale stocks and save the industry.

Salvesen was founded as a business company in Scotland in 1846. The Salvesens were from Mandal, a town on the southern tip of Norway, which during the nineteenth century was one of the country's principal sailing-ship ports. They were a merchant family which became modestly prosperous in the early years of the century through timber exporting and shipping. One of them, Ole Salvesen, diversifying from his commercial activities, fitted out a boat as a privateer. This was encouraged by the Danish government and several Mandal owners followed suit; but none of the ships succeeded in taking any prizes and the idea was soon abandoned.

Ole's son Thomas had a family of six sons and three daughters. There was not enough in the family business to absorb so many, so one son, Theodor, moved to Scotland, followed a few years later by another, Christian. Theodor set up at Grangemouth, at the mouth of the Forth-Clyde canal, where there seemed to be burgeoning opportunities. Christian, who was my great-grandfather, opened an office in Leith in 1853, initially as a branch of Theodor's firm, but as the brothers developed separate interests they moved apart and in 1871, after Theodor's early death and the dissolution of his own business partnership, Christian became independent.

Christian Salvesen & Co.'s interests spread from the initial ships' agency into shipowning, timber and coal, keeping close links with the family interests in Mandal. The company became brokers for the sale of Norwegian whale oil in Scotland and supplied coal cargoes to whaling stations. That brought them close to the Norwegian whaling industry, which was already interesting their kinsmen in Mandal.

Eventually, in the late 1890s, Salvesen bought a whaling station in Iceland, extending then to the Faroes and Shetland. These new ventures were led by Christian's son Theodore, my grandfather, who with his elder brothers Tom and Fred had joined Christian in the partnership.

Salvesen came into Antarctic whaling in 1907, when *Coronda* sailed from Leith to set up a whaling station at New Island in the Falklands. A year later they extended to South Georgia, further south, to found the station of Leith Harbour, which became the centre of our southern whaling ventures and continued until the sale of our last floating factory, *Southern Harvester*, in 1963. The span, therefore, of our Antarctic activities covered little more than the working life of a single generation. When I was first in the Antarctic in 1948 I could meet people who had been with the first whaling crews who set up Leith Harbour. Our leading gunner, Hans Abrahamsen, star of the expeditions from 1927 to 1960, started his whaling life as a whalecatcher fireman off Spitzbergen in 1909 and came to the Falklands in 1915.

The trails that led to the opening of the Antarctic by Norwegian whalers started with Svend Foyn, the Norwegian sealing captain who in the 1860s and '70s transformed the traditional whaling industry. For centuries men had hunted whales with rowing boats and hand harpoons. As they looked for whales further from their shores they equipped ships to carry the catching boats and their crews and to process the whales. The earliest distant whaling grounds were in the Arctic, round Greenland and Spitzbergen, where the Dutch set up a thriving industry in the seventeenth century and the British, somewhat less successfully, followed. The Americans, meanwhile, took up the trade from Massachusetts. Their ships ranged the oceans of the world, sometimes not returning to their home ports of Gloucester, New Bedford or Nantucket for three years. Whaling on this pattern was an arduous and dangerous pursuit, as chronicled in the logs of many American whalers and in the pages of Herman Melville's *Moby Dick*.

The ships were fitted out to process and store the products of the whale. The main prize in the Arctic was the right whale. Its long baleen plates (the flexible fibrous strips through which the whale sieves its feed of krill or fish) were highly valued for the stiffening of corsets. Blubber was also taken, rendered down to oil by cooking pots on deck, or stowed in chunks in the hold. The Americans chiefly looked for the sperm whale, abundant in the tropical waters of the Pacific, but also took right whales and humpbacks in the colder waters north and south. The oil from whales was sold for soap, lighting, lubrication and industrial uses, though the development of the mineral oil industry in the USA and the growing use of oils from crops such as linseed and cottonseed put steady pressure on prices.

The heyday of American whaling was the first half of the nineteenth century. By the later years of the century there was little left but romance. Whales became scarce, ships came back from long cruises with empty holds and crews went to seek their fortunes elsewhere.

The primitive techniques of the whalers of that era limited their catch to the slow-moving whales which could be approached and harpooned by muscle power. They also had to float at death. Sperm, right, and humpback whale had those qualities, but there were huge populations of baleen whales, headed by the blue and fin, which could not be touched. It was these whales, abundant off the coast of Norway, that Svend Foyn was resolved to catch and for over twenty years, while still operating a very successful sealing business, he single-mindedly pursued his objective. Progress was slow as he gradually improved his equipment and tested it in the storms of the North Sea. Eventually he developed a steam whale catcher, which could chase, kill, and recover the blue and fin whale. It had a heavy harpoon gun at the bow which fired a harpoon with an explosive grenade head. The harpoon line passed through a system of pulleys and springs which absorbed the shock of the harpooned and diving whale and was controlled and wound in by a steam winch. Svend Foyn's whale catcher was the foundation of the modern whaling industry. Over the years the ships grew enormously in size, power, and refinement, but the basic model remained the same.

Svend Foyn was inspired in his labours as much by the conviction that he was doing God's work in harvesting the seas as by the vision of commercial success. His diary, which chronicles his labours, intersperses his accounts of failure and triumph with prayers to the Almighty to support him to a fruitful outcome. His home town, Tønsberg, on Oslofjord, with the neighbouring towns of Sandefjord and Larvik, became the base for the new industry. Svend Foyn is still honoured in Tønsberg; his statue is crowned in the National Day celebrations of 17 May and a speech is delivered in his memory.

It was a giant step for the whalers to take from the confined seas of the north to the vast oceans of the Antarctic. But the first decisive move south, in 1904, was not a blind one, since there had been considerable reconnoitring activity over the last decade of the nineteenth century. The old whaling firms of the Arctic, now on their last legs, hoped that the Antarctic might provide catches of right whales which would keep them going for a few years longer. The Norwegian whalers, with Svend Foyn's modern equipment, were already finding that whales in the north were scarcer. They saw in the south a far bigger sea area which might yield an abundance of blue and fin whales.

In 1892 two expeditions set out for the Antarctic. One was led by the Norwegian *Jason*, captained by C. A. Larsen, later to become the founding entrepreneur of Antarctic whaling. The other, from Dundee, was headed by *Balaena*. Both were old-style ships, working with rowing boats, and their mission was to find right whales and seals. On *Balaena* there sailed two young men, Dr W. S. Bruce, who later became a noted Antarctic explorer, and W. Burn Murdoch. Burn Murdoch recorded the voyage of *Balaena* in *From Edinburgh to the Antarctic* illustrated by his own drawings. He took part later in sealing voyages to the Arctic, and even tried to start his own whaling station in the Azores in 1913, with a diesel-powered whale catcher. Several of his pictures were hung in the club house at Edinburgh Zoo, one showing *Balaena* as a tiny ship among towering tabular icebergs. Harold Salvesen, of the next generation of Salvesen whalers, was critical of the size of the icebergs and characteristically passed on his criticisms to the artist. Burn Murdoch dismissed him as having no sense of perspective.

The reports brought back by *Jason* and *Balaena* and the expeditions that went out in the following years were that there were plenty of seal in the ice in the Weddell Sea, few right whales, but apparently good stocks of fin and blue.

The Scots on their return tried to interest Edinburgh businessmen, among them the Salvesens, in starting a station at South Georgia, but at that point Christian and Theodore were sceptical. They no doubt accepted that the whales were there to be caught, but the difficulties of setting up and running a station on a bleak, uninhabited island at the other end of the world, with ships, men and supplies all to be sent from Norway or Scotland, and products to be shipped back to Europe for sale, seemed formidable. The Norwegians were similarly not persuaded by C. A. Larsen's enthusiasm.

The Antarctic was now becoming a popular area for exploration, one of the world's last frontiers. There was *Discovery*, with Captain Scott, Gerlache's Belgian expedition, German and French ventures, Bruce's Scottish National Antarctic expedition, and finally the Swedish South Polar expedition under Nordenskjold with C. A. Larsen, appropriately, chosen as master of the *Antarctic*, Nordenskjold's ship. *Antarctic* got caught in the ice in 1903 and Larsen and her crew had to abandon her. They were eventually picked up by an Argentine ship and landed at Buenos Aires. It was there that Larsen found support for his idea of whaling at South Georgia. Capital was raised, and in November 1904 the first station was founded at Grytviken – the 'bay of pots', so called after the big iron try pots left behind there by early nineteenth-century sealers and still to be seen flanking the old manager's villa.

Not far behind Larsen came Christen Christensen, who had commissioned the earlier *Jason* expeditions. He had been operating a floating factory, *Admiralen*, off Spitzbergen, working with two or three whalecatchers which kept it supplied with whales. He took the bold step in 1905 of sending that fleet to the Antarctic. *Admiralen* started off in the Falklands at New Island and proceeded from there to Admiralty Bay in the South Shetlands. Though the venture barely broke even, it made a reasonable catch and showed that Antarctic whaling was feasible and could be profitable.

Salvesen was only one of several companies which opened stations at South Georgia in the wake of Larsen's 'Compañia Argentina de Pesca' at Grytviken. After some initial hesitation the British government, represented by the governor of the Falkland Islands, took control of whaling both at South Georgia and further south in the South Shetlands and Graham Land. The Colonial Office licences limited both the number of expeditions and the number of whalecatchers to operate with each station or floating factory. The first concessions were given without operating conditions, but later the licences for land stations required the full working up of the whales. This applied to Leith Harbour, and the initial shipment of whaling equipment to South Georgia included a guano plant to process and dry the bones and meat for fertiliser after the extraction of the oil. The stipulation was impracticable for floating factories, which at that time were often too small to do more than cook out the blubber, leaving the carcasses to rot or be processed by an auxiliary factory ship.

The Salvesen station at New Island in the Falklands did reasonably well, although it was too far north and also away from the main movement of whales to and from the Antarctic. But the station at Leith Harbour found an abundance of whales, initially humpbacks, later blues and fins, and prospered from the beginning. Four whalecatchers worked for New Island, all brought from the Shetland station and returning there for the summer fishing. Leith Harbour started with only two boats, but brought in two more the second season. The first whalecatchers had been built at the Akers yard in Oslo, but the new ones sent down for the 1909/10 season were from Hawthorns, a little yard on the Water of Leith not far from the Salvesen office. They were 95 feet long, with a speed of 11 knots. The Hawthorns whalecatchers continued to serve the company until 1924, when they were superseded by bigger more powerful catchers and sold. One of them, *Sonja*, went to Greenland. She was a popular ship as every whale she caught meant winter food and warmth for the local Eskimos. There was a special song, 'Sonja has whale', which was sung when she delivered her catch.

By 1913 there were ten stations or floating factories working at South Georgia and twelve floating factories in the harbours of the South Shetlands, the South Orkneys and Graham Land. The South Georgia stations were all set up on the north coast of the island, where there were deep fjords and the mountains gave some protection from the prevailing gales. Grytviken, and later the government colony of King Edward Cove, were sited in Cumberland Bay, the biggest inlet on the coast. Stromness Bay, however, a few miles further north, was found to be a more suitable area, and three stations, Leith Harbour, Stromness and Husvik, were built in adjacent arms. The first two had the great benefit of deep water right up to the shore, so the large floating factories and transports of later generations could tie up at their quays.

In the South Shetlands and further south the factory ships found harbours with shelter to provide calm conditions for flensing the whales (stripping off the blubber) alongside and fresh water to be had for the cooking processes. Deception Island in the South Shetlands became for twenty years the centre point of Antarctic whaling, its drowned crater providing a magnificent enclosed harbour. The floating factories congregated there, with their attendant whalecatchers, some of them moving south along the Graham Land coast when the ice pack receded and returning in the autumn as the ice closed in. It still keeps some volcanic activity; hot springs and steam rise through the sand on the shore, and on at least one occasion in whaling history the bottom has suddenly moved, leaving the ships with their anchors dangling. By the 1930s the whalers had departed, leaving a memorial of bleached whale bones along the shores. After the Second War it was recolonised by scientific stations from the UK, Argentina and Chile, which indulged in occasional war games between themselves. Whalers only came back in 1955 after a period when the area had been closed for whaling.

Graham Land was so christened by the British, while the Americans have insisted on Palmer Peninsula. To avoid controversy it is currently referred to as the Antarctic Peninsula. Our operations have left their mark on the maps – at Neko Harbour, Salvesen Cove and Theodore Island.

The Colonial Office restricted its licences for stations and floating factories and their catchers. It also levied a tax on the production of each expedition. This régime continued throughout the whaling period for the South Georgia stations. Floating factories eventually escaped from it when in the late 1920s they left the Antarctic harbours and did their fishing in the open seas. The restriction of licences was intended to give some protection to the whale stocks, though no one had any idea of their extent. The governor was continually resisting pressure from the whalers to issue more licences; clearly the limits imposed by government prevented even greater exploitation of the whale stocks. The number of expeditions allowed to fish in the area, after a very rapid expansion between 1909 and 1913, stabilised at twenty-two. Those who subsequently wanted to join had to buy a licence from one of the existing holders.

Our company's relations with the governor were not always cordial. Theodore wanted as many licences as he could get, feeling that Salvesen as a British company of high reputation should as a matter of course be treated more favourably than its Norwegian competitors. The civil servants saw it as their duty to be even-handed between the many companies distastefully clamouring for the opportunity to make money out of the Antarctic. They were sceptical about the patriotic argument when most of the Salvesen crews and much of the equipment came from Norway, and they did not much like the way in which Salvesen managed to make their interests coincide with the highest moral principles, a line passed on to Harold Salvesen in the next generation. But Theodore was in harmony with government in his enthusiasm for the full utilisation of the whale, which had been standard on Salvesen northern stations. Most of the Norwegian entrepreneurs, with scanty capital and poor equipment, could at first see profit only in taking as many whales as possible, stripping the blubber and dumping the rest. They soon came to realise, particularly after the cream of the local whale stocks had been taken, that they could not afford to let half the available oil go. In 1913 government pushed proper utilisation further by prescribing a minimum processing capacity of pressure cookers on floating factories. But the old habits of strip and dump still persisted at times of high catch.

The floating factories of that era were old ships bought and converted. They were quite often passenger ships because these had extra accommodation and could provide space for the factory equipment without the expense and risk to stability of putting on an extra deck. The whales were tethered alongside for flensing. A Norwegian businessman, who at the age of fifteen had sailed with our *Sevilla*, explained to me forty years later the perils of being a 'jolle-gutt'. He was the whaler who stood on a boat moored to the outside of the carcass trying to keep the whale steady with a boat hook while the flensers climbed over it to remove the blubber. He was soundly cursed when the whale rolled and his arms ached to the limit of endurance. The strips of blubber were hoisted aboard by derrick, cut into smaller pieces and fed into cookers, originally open, but soon closed to give extraction by steam pressure. The oil was tapped into barrels, while the remains of the blubber were put overboard. The original steam engines were kept for driving the ships. The steam from the boilers could operate the engines or run the factory, but not both together. It was only when open sea fishing started at the end of the 1920s that sufficient steam power was needed to allow a floating factory to process and move at the same time. While the factories remained in harbour at anchor that requirement did not arise.

The factory needed plenty of fresh water for the cooking process and to make up steam losses. The ships had fresh water tanks for the engine room, and sometimes evaporators, but these were often inadequate, so the factory ships had to find bays where there was fresh water available, and carried water boats which ferried the water between shore and ship. Some of the harbours at South Georgia still have dams on the hillside put up to provide water for the floating factories that anchored there. The development of open sea fishing took the ships away from that source of supply, so the factory ships had to carry more water and fit more evaporating capacity. Factory ships also generally had to have enough coal storage capacity for their own needs and that of their catchers over the season, though from the late 1920s, when coal was replaced by fuel oil, new fuel supplies could be brought in during the season by tanker and pumped over to the factory.

Salvesen's first floating factory was *Neko*, a twenty-year-old steamship of 3,500 tons. She joined the South Shetlands fleet in 1911 with four catchers. With her was bought *Horatio*, of similar size and age. It was intended that *Horatio* should cook out the carcasses that *Neko* or Leith Harbour could not cope with. *Horatio* had an unfortunate career, with an initial fire followed by an unsuccessful foray to Africa, eventually sinking after a spectacular fire at Leith Harbour in 1916. *Neko* did well, joining the Deception and Graham Land fleet from 1911.

In 1910 Salvesen had become the biggest whaling company in the world. It had five stations in Iceland, Faroes, Shetland, Falklands and South Georgia and with twenty whalecatchers took 2,350 whales and produced 66,500 barrels of oil, i.e. 11,100 tons. (The traditional measurement of oil production by barrels continued to be used in modern whaling. There were six barrels to the ton.) This however was still an industry in its infancy. Salvesen's total production then was tiny compared to the 3,600,000 barrels produced from the Antarctic at the height of the industry in 1931 and to Salvesen's own production record of over 350,000 barrels in the period after the Second World War. Correspondingly our employment of 400 whalers in 1914 contrasts with over 2,000 in the post-1945 expeditions.

The original Leith Harbour station, like other whaling stations of the time, was a simple affair. There was a slipway to haul up the whales, a wide wooden deck (the plan) on which the carcasses were stripped of blubber and cut up, steam cookers to cook out the oil, and a guano plant, a long horizontal rotating cylinder which tumbled the cooked meat and bone through a direct flame to dry it into meal. There was a boiler house providing steam for the factory and for generating electricity. A bit away from the factory were the wooden barracks and mess rooms for the whalers, and the manager's villa and office. As time went on storehouses and workshops were added and a slip and workshops for hauling up catchers and repairing them. The station was built close under Mount Coronda, the 1,700 ft. peak rising steeply out of the bay, probably because it was the flattest and most sheltered site, but it was cramped, and exposed to winter avalanches. As the station expanded it was moved down the bay and finally the original site became the station scrap heap, appropriately named Jericho. Jericho was a useful mooring area for laid up whalecatchers, and also accommodated *James W. Turpie*, an iron steamship from 1881 brought down to be a coal storage hulk. When oil superseded coal as fuel for the station and catchers she became redundant and eventually sank at her moorings, with her side plates just visible on the surface at low tide.

The man who led the whaling ventures in the Antarctic was Henrik Henriksen, an engineer from Tønsberg, who had already supervised Salvesen's operations in the north. It requires someone of no ordinary qualities to set up a whaling station, with all its cookers, tanks, boilers and living quarters, on a bare beach at the end of the world and make it work within a few months. Henriksen was clearly outstanding. He belonged to a family of blacksmiths which had designed harpoons and guns for Svend Foyn. His father and brother prepared the layout of the Leith Harbour station and supplied the cooking equipment from their workshop. Henrik was manager of the station from the start until his retirement in 1921. The firm of Henriksens Mekanisk Verksted still flourishes under a great-nephew of Henrik, and has among its mementoes a pair of beautiful duelling pistols, by Purdey of London, in a handsome oak box, presented by the Salvesen partners to Henrik Henriksen on his retirement.

The new factory was originally known as Jedder, after a hard labour installation for criminals in South Norway. That name was dropped, presumably as the station acquired more amenities. One valuable addition was the cinema, a welcome release for whalers from the monotonous grind of processing whales or humping stores.

Most of the whalers were Norwegians recruited from around Tønsberg. A few Shetlanders were sent down from the Olna (Shetland) station, but they found life difficult among the Norwegians and did not return for a second season. The floating factories had British masters, as required by United Kingdom law, and the transport ships ran with British merchant navy crews. But there were also periods, particularly during the First World War, when the company had to grab anyone it could get. Some were brought to the station from

Buenos Aires and from St Vincent in the Cape Verde islands, which was a coaling call for ships on the way south. The islanders, being unskilled, were usually put into the coal gang, the least attractive job in the station. This extended to the transport ships. Leganger Hansen, the free-spoken manager of Leith Harbour, wrote home on one occasion: 'I have seen some curious crews in my time here, but nothing like the gang on that ship. Said to be 17 different nationalities, and that is lumping all the blacks together as one. I said to the captain that he had a fine crew. He replied that it was the worst pile of shit he had ever had.'

The practice of recruiting for the station on the hoof came to an end in the 1920s, but there was always a problem of stowaways from Cape Verde and the Canaries. They found it easy to hide in the stacks of empty barrels that were carried south on the decks of the cargo ships to be filled with whale oil. They popped out, sometimes in scores, when the ship was well clear of St Vincent. The master then had to choose whether to turn back and drop them ashore or continue with them to South Georgia. He usually did the latter. When they arrived at Leith Harbour the manager gave them work without pay for the season but made them full whalers if they stayed on for the winter and beyond.

The outbreak of the First War pulled the whaling industry out of incipient depression. The increases in production of whale oil had been too much for the fairly static market for soap and lubricants, and margarine, which was to be the predominant use for whale oil, had not yet caught on. But war brought the need for glycerine, important in high explosives, and increased demand in other areas. Encouraged by government, catching effort and production were greatly expanded at Leith Harbour. In 1917/18 there were eleven boats fishing from the station and a production of 94,000 barrels, compared with the 1913/14 figure of 37,000. The productive capacity of the station was expanded, partly by dismantling the New Island station, which was no longer worth continuing, and bringing its equipment to South Georgia. But that was not enough to cope with the increased catch. To ensure that the maximum oil was produced for the national need the British government suspended the requirement of full utilisation. Bones piled up on the shores round the station.

Catch at South Georgia depended at first on the huge stocks of humpback whales found there. In the 1910/11 season over 6,000 humpbacks were taken. They were however very quickly fished out. By the 1917/18 season the catch had gone right down to 60, and fishing had passed to the still abundant blue and fin whales.

The problems of shipping men and stores down to the station and getting the whale produce home through U-boat infested seas were considerable. *Coronda*, which had pioneered Falklands and South Georgia, was sunk by torpedo, as were other supply ships. Coal and barrels became scarce and expensive. But the Norwegian stations, without the help of a government at war, fared even worse, and their catches lagged well behind Salvesen's.

It was during the war, in 1916, that Ernest Shackleton came to South Georgia from Elephant Island where he had been shipwrecked. His crossing of the South Shetland Sea through the winter storms in the lifeboat *James Caird* is the ultimate epic of Antarctic exploration. He and his companions landed at King Haakon Bay, on the opposite side of the island to the whaling stations. In a thirty-six-hour march they traversed the mountains of South Georgia, finally emerging on the col above Stromness. To their immense relief they saw far below a catcher steaming out of the station and heard the hooter sounding for breakfast. They had feared that the station would have closed for the winter and all the whalers gone home. The manager at Stromness, Sørlie, received them warmly, sent a catcher round to King Haakon Bay to take off Shackleton's two companions left there, and arranged to get Shackleton to Buenos Aires to rescue the remainder of his crew from Elephant Island. Stromness station was next door to Leith Harbour. It eventually stopped operating in 1931, and we later took it over for the drydocking and repair of the catcher fleets.

South Georgia has never attracted any settlers beyond the whalers and government officials and scientists. It might have been different if the island had been on the right side of the Antarctic convergence, which marks the north boundary of the cold waters. Tierra del Fuego has abundant grass, trees, even strawberries. South Georgia, at the same latitude but south of the convergence, has little beyond the coarse tussock grass and mosses. Penguins and seals, feeding on the rich fish and krill, thrive and multiply, but there are no native land animals, and those that have been imported are hard put to survive. The pasturing of sheep, goats, and cattle to feed the stations was tried, but they were unable to stand the long winter snow cover and the scanty grazing. Reindeer, imported from

Norway, have done better, though over the years they have bred to be considerably smaller than their parent stocks. But they provided useful fresh meat, and the stations used to organise regular reindeer shoots. Below ground there were hopes of minerals. Theodore Salvesen sent a geologist down in 1912 to survey the island, but he found neither oil, gold nor coal. The riches of the island lie in the whales, krill and fish that surround it. Later generations may find a different satisfaction in the striking landscape of snow-covered mountains, the vast penguin colonies, and the elephant and fur seals covering the beaches, while the decaying remains of our whaling stations add archaeological attraction.

The living conditions of the Leith Harbour whalers of the time were far from grand, though probably no worse than normal in the merchant navy. The transport ships had the fore peak or tween deck areas fitted out with tiers of wooden bunks. Our Norwegian anecdotal history has a vivid account of the voyage south from Toralf Fritzvold, baker at Leith Harbour for thirty-three seasons. In bad weather the hatches were battened down, leaving the tween decks without daylight. The space sometimes had to be shared with the odd sheep and pig which had survived being swept overboard from the deck pens in the angry forties. The station personnel could count on a dietary routine of meat – salt and tinned – supplemented by excellent whale meat and the local fish. Potatoes sometimes ran out, but there was always fresh bread. Penguin and albatross eggs, too, were collected regularly. To combat the risk of scurvy – of which there were a few cases – one of the early doctors got the whalers to eat the roots and young leaves of tussock grass.

After the end of the war it took several years before the levels of whaling reached their prewar peaks. The Norwegians had diverted many of their floating factories into the cargo trade, and the high prices of the immediate postwar boom discouraged any large investment in new ships. Salvesen continued with *Neko* in the South Shetlands and with the Leith Harbour station. *Neko* was joined in 1922/23 by *Sevilla*, a much larger ship, converted from a German passenger liner. *Neko* was lost in May 1924 close to Rio de Janeiro on her way home from the whaling season. Her replacement, about the same size as *Sevilla*, was *Saragossa*, which first operated in 1925/26. The catcher fleet was progressively modernised during the period. The new catchers, now all built by Smith's Dock at their South Bank yard in Middlesbrough, were between 110 and 130 feet long and up to over 1,000 h.p. The larger of the new boats had been designed specifically for Leith Harbour. Whales could now only be found fifty miles or more out from the station and more horsepower was needed to reduce the time for towing in.

Our own progress was paralleled by the Norwegian companies, who consolidated their dominant position in the Antarctic with bigger floating factories, among them *Ronald*, the first floating factory built as such.

Whaling was stimulated in the interwar period by the development of new markets for whale oil. A process for hardening and purifying oil by hydrogenation had been patented in 1905, and as it was developed whale oil became more attractive as a principal raw material for soap and margarine. Sperm whale oil, which has a different composition from baleen whale oil, being more akin to a wax, could not be used for edible purposes but found a separate market in special lubricants and cosmetics.

The twenty years between 1905 and 1925 saw an explosive growth in whaling activity throughout the world, not just the Antarctic. This was almost totally led by Norwegian companies. It encompassed Africa, from the Cape to the Congo, Australia, South America, Mexico, Alaska, Labrador, Korea and Japan. Japanese companies built up their own industry, helped initially by Norwegian gunners and equipment. Many of these new whaling ventures had short lives. The whales, often humpbacks, were easily caught, but the stocks were quickly exhausted. Whaling in tropical waters is anyway difficult to make profitable, as whales carry a much thinner layer of blubber there than in the colder waters where they feed, so their yield of oil is poor. Although a number of these far-flung enterprises lost their capital, the Norwegian whalers showed remarkable energy and initiative in pursuing their calling all over the globe. Salvesen, well established in the Antarctic and still working in Shetland, did not extend further except to take over a station at Saldanha Bay, near Cape Town, and – much later – a small operation at Hawkes Bay in Labrador.

After the end of the war the Colonial Office régime of restricting licences and controlling utilisation of the whale was reintroduced. There were many applications for new licences, but the British government stood firm in not extending the number. It was equally firm in resisting the Salvesen claim that in any further grant of licences British companies should have preference, well aware, from the experience of North Sea whaling, that the seemingly abundant stocks of Antarctic whales off South Georgia and the South Shetlands could easily be exhausted. In 1921 the catching of humpbacks, which had formed the bulk of the catch in the first years at South Georgia, was banned, while from 1919 part of the revenue levied by the government on whaling activities was set aside for scientific work. There was still little known about the life cycle and reproductive patterns of the various types of whales and of their migratory movements between the cold Antarctic seas which were their summer pastures and the warmer seas to the north, where they trekked for the winter. Such knowledge was essential to provide a proper base for assessment of whale stocks and regulations for their protection. This work was organised through the Discovery Committee, which bought and operated Captain Scott's ship *Discovery* and *William Scoresby*. The ships covered the whole Antarctic making surveys on whales, mainly through the simple technique of counting whale spottings. They also studied sea currents and temperatures, plankton, and the whole complicated structure of ocean life which culminates in the Antarctic's abundance of fish, seals, and whales.

The Discovery Committee was not popular with the Norwegian whalers, who saw its activities as an attempt to promote British interests at the expense of the Norwegians. It was difficult for them, however, to make any convincing protest, as they had been treated with scrupulous fairness in the issuing of licences by the British government. In any case the Norwegian whaling industry, still dominant in the Antarctic, was now coming under fierce public attack for its massive and wasteful slaughter of whales. Harold Salvesen was also later extremely critical of the *Discovery* operations as he had come across them in the Antarctic, and blasted the government for its lack of control: 'The whole business has been conducted most inefficiently and at criminal cost.'

The Colonial Office régime was reasonably effective in regulating whaling in the area under its authority. Apart from limiting the numbers of land stations, floating factories and whalecatchers, and trying to ensure full utilisation of the whale, it restricted the catching season at South Georgia, where the stations, unhampered by the winter ice which stopped operations further south, could otherwise happily operate throughout the year. Except for the complete protection of the humpbacks and the ban on whale calves or mothers with calves, the Colonial Office did not try to control the numbers of whales taken. Catches continued high up to the mid 1920s, and there were no strong grounds for supposing that the stocks were being grossly overfished. In any case it was already clear that the whales which appeared round South Georgia and the Antarctic peninsula were only a small part of huge populations spreading right round the oceans of the Antarctic continent.

This period of British government control did not last long. Whaling, when it had settled down after the upsets of the First War, was a profitable business, indeed for some very profitable. It had become an important part of the Norwegian economy. Companies who had been unable to get licences from the Colonial Office, or who wanted to expand their operation beyond the limits of their licences, looked beyond the three-mile coastal limit controlled by Britain to the open seas, where they could fish without restriction with the facility of the larger floating factories and more powerful catchers which were being introduced. They now had sufficient power to get out of trouble if the ice bays in which the factory ships worked started to close in. The supplies of fresh water needed for steam raising and processing could be carried by the ship or evaporated from sea water on board. The constant danger of damage or sinking by ice was also declining as the ships became bigger and better strengthened. The introduction, too, of direction finders and wireless was diminishing another hazard – fog, particularly frequent in the Weddell Sea. Catchers could go far afield after whales and be unable to find the factory ship again.

The main technical obstacle to open sea fishing, or ice fishing as it came to be called, was the need to cut up the whale in the sea alongside the factory ship and hoist the pieces up to the deck. This was a laborious process and required calm waters. A floating factory could often find such conditions within the ice pack but had to suspend operations whenever sea movement developed. In the mid 1920s a triumphant answer was found in the development of the stern slip, a ramp by which the whole whale could be hauled up tail first on to the deck, to be cut up there in safety. The Norwegian *Lancing* was the first factory ship to be fitted in 1925, but there was no rush to put slips on existing factory ships. Some of them were not suitable, and there were problems of gear and winches which took some time to solve. But the slip quickly became standard for any new additions to the Antarctic fleet. Now the floating factory could not only haul up its whales, flense them and cut them up for processing regardless of weather, it could also move with the catchers, unhampered by outboard operations.

The shift away from coast fishing to sea fishing was gradual, partly because the coastal catches continued to be good. But a poor season in 1926/27, with a scarcity of whales on the normal trekking routes combined with bad weather, convinced most of the whalers that the future lay in ice fishing. By the 1928/29 season the traditional harbours of the South Shetlands were almost deserted, and 80% of the total Antarctic catch that season was taken in the ice.

The ice fishing started in the Weddell Sea, close to the South Shetlands fields, and, on the other side of the world, in the Ross Sea, where the South Georgia pioneer C. A. Larsen took a first expedition in 1922/23. It soon covered the whole Antarctic round the globe. Lars Christensen, one of the biggest whaling operators, extended the exploration of the Antarctic by a series of expeditions sent out between 1927 and 1931. The aim was to look for new whaling grounds and to find new lands for Norway which might be used as bases for Norwegian whaling. His ship *Norvegia* investigated, and annexed for Norway, Bouvet Island, south south west from Cape Town, and Peter I Island, west of the Antarctic Peninsula, as well as part of the continental coast line. These islands were specks of volcanic rock in stormy seas which gave no haven for whaling or any other activity. The cruises, which covered the whole Antarctic round the continent, disposed of a number of supposed islands which had been sighted over the years and entered tentatively on charts, but which were no doubt large icebergs which had broken off from the continental ice shelf and been taken for land. The whalers came across such floating ice islands from time to time, up to 100 miles long. One of our tankers on the way in to South Georgia thought it had picked up the coast line on its radar but found it was a gigantic iceberg fifty miles out.

The most important contribution by Lars Christensen to the whaling industry was the discovery of new and abundant whaling grounds round Bouvet and further east, north of Enderby Land. Both these areas became popular fields for the floating factories, though Enderby was also notorious for storms. The only area where whales seemed to be scarce was the Bellingshausen Sea, west of Graham Land, as far as Peter I. That area was later made a whale sanctuary by international agreement.

The new prospects thus opened up, combined with rising prices for whale oil, created a massive wave of optimism in Norway. New floating factory expeditions were commissioned and new companies launched, often with British capital behind them. Between 1927 and 1931 the Antarctic fleet increased dramatically. In 1927/28 there were 17 floating factories with 61 catchers; in 1930/31 this had risen to 41 floating factories and 200 catchers, while seasonal production had grown from 733,000 barrels of whale oil to 3,400,000.

Our company, with its large station at Leith Harbour and licences for two floating factories, had a strong position within the licensing régime and kept continual pressure on the Colonial Office to improve it. But as ice fishing developed Salvesen too adapted its floating factory operations. *Sevilla* was sold and replaced by *Salvestria* and *Sourabaya*, much larger ships with stern slips and high processing capacity. The new generation of floating factories, ranging over the Antarctic, brought in much larger catches than the old harbour-bound expeditions. *Sevilla* and *Saragossa* were pleased with production of 50,000 barrels each. *Salvestria* and her sisters regularly achieved 100,000.

The whaling side of the company had been started and developed by Theodore Salvesen, with great vigour and success. By 1918, the end of the Great War, he was in his late fifties and strained in health. It was not long before he brought in his three sons, first Noel, then Harold and Norman, to take on the firm's management. When they were joined by Iver, the youngest son of Tom, all the surviving sons of Tom and Theodore had been brought into the family firm. Tom's two elder sons had been killed in the war, and Fred had no children. A further son of Christian, Edward, became an advocate and a judge of the Court of Session. One of his grandsons, my cousin Maxwell Harper Gow, entered the firm after the Second War as the first of the next generation. Iver followed his father on the merchant shipping side of the company, while Norman concentrated on the ships' agency and chartering business.

Theodore had married into an intellectual family. His wife, Annie Burnet, was one of the first women graduates of London University and sister of the distinguished St Andrews Professor of Greek, philosopher and translator of Plato, John Burnet. This trait came out in Noel and Harold. Both of them had strong academic interests which strengthened their business careers. Noel was perhaps a reluctant businessman, as he had originally set his heart on the law. But in the late 1920s and up to the outbreak of war in 1939, when he decided to take on military service, he had, under Theodore, a leading part in the whaling side of the company. There were two particular issues over which his legal mind showed its worth. One was in defeating a claim by Unilever to reject Salvesen's whale oil production on grounds of fuel oil contamination. That was a pretty blatant attempt by Unilever to escape the consequences of a severe fall in whale oil prices after the sale contract had been made. The second achievement was to persuade the Inland Revenue commissioners that they should not surtax the company for retaining its earnings rather than paying them out as dividends. Private companies which wanted to lay up reserves were always under suspicion by the tax authorities that the owners were primarily trying to avoid the tax that would be levied on the individual shareholders if the money were paid out. This argument continued for many years while the company was owned by a few large family shareholders. Fortunately Noel, and Harold after him, was able to maintain that the fluctuating fortunes of whaling required a substantial level of company reserves. The policy of small dividends and large retentions built up the company and was of immense value to its development after the Second War.

Harold arrived in the company via the Indian Army and Oxford. Having taken a first class degree at Oxford in the new school of politics, philosophy and economics, he stayed on as an economics lecturer at New College, but was of far too ambitious and aggressive a temperament to be happy for long as a university lecturer. After three years, in 1928, at the age of thirty, he moved into business and joined the Salvesen firm. His first job after he had learned Norwegian was to go to the Antarctic and study company operations with a view to future policy. Combining his formidable intellect with a vast appetite for work, he immersed himself in whaling. He spent the whole 1928/29 season in the Antarctic and went down again the following year with the management of the company operations in his hands.

Harold came home from his first season captivated by whaling and enthusiastic about its prospects. This overcame Theodore's doubts about the whale stocks and his worries on whether a heavy new investment in floating factories and catchers could pay for itself before the whales ran out. Harold, under the experienced supervision of Noel, became closely involved in the conversion of the two liners, *Cardiganshire* and *Carmarthenshire*, fifteen and thirteen years old, which became the floating factories *Salvestria* and *Sourabaya*.

During his Antarctic trip he had visited several of the Norwegian factory ships and had been struck by the slackness of management of their operations and the inefficiency of their processing. So he interested himself particularly in getting the best possible equipment into the new ships. *Salvestria* and *Sourabaya* came out with processing capacity a good way ahead of their competitors.

The Norwegian whaling pattern, unchanged since the pioneer days, was simply to catch all the whales they could. How they were processed was a secondary matter. As long as there were plenty of whales coming in it did not matter much if you only processed the blubber and some of the bones. Their philosophy was expressed to Harold by a hoary Norwegian whaling manager: 'Catching is the same the world over. You must grab what you can before the next man comes.' The Norwegian entrepreneurs hired the best gunners they could find, took on an experienced, and preferably sober, captain for the factory ship and left them to get on with it. The Norwegian managers in the Antarctic accepted that and resented interference in their sphere from the office at home. Tom Humphreys, who later became a partner in Salvesen, was sent down to South Georgia as a young man by Unilever to inspect their station at Prins Olav Harbour, up the coast from Leith Harbour. The Norwegian manager there refused to let him land, and he had to stay kicking his heels on the ship until a telegram was sent from home instructing the manager to receive him.

The organisation of the Salvesen expeditions was initially not much different from this, though Theodore's interest from the beginning in getting proper utilisation of the whales set its mark on the floating factories as well as the land stations. Most of the detailed management of the whaling fleet was done through the great Leganger Hansen, manager of Leith Harbour between 1921 and 1937. Neither Theodore nor Noel ever visited the Antarctic whaling grounds, nor did any of the Norwegian owners, except Lars Christensen on his survey cruises. With this gulf between the home directors and the whaling managers in the Antarctic it was not surprising that Noel on one occasion sent a telegram to the gunner manager of one of the expeditions on the lines 'Very disappointed at your poor catch', and got the reply 'Jeg au' ('Me too'). Harold changed this. Having made himself an expert on whaling and seen the gross deficiencies of whaling practice at first hand, he started to manage Leith Harbour and the floating factory operations as a modern industrial concern. The standards he set on planning and technical expertise, and the friendships he made with whalers at all levels, brought to the company a quality and pride which carried on to the end of whaling and beyond.

Harold in his youth did not always hit it off with the Norwegian whalers. On one occasion a party of flensers, goaded no doubt by his insistent questioning, formed a circle round him and brought down their flensing knives together within a few inches of where he stood. Harold kept his ground and they dispersed with some respect.

Harold kept his military title of captain. He was referred to in Norway and throughout the fleet as Captain Salvesen, in the Leith office as Captain Harold. Such claim of rank was more common then than now, and seemed to be particularly favoured by the Salvesen partners. Theodore was often addressed as Colonel Salvesen, having commanded a territorial unit of engineers, and Noel sometimes became Major Noel. They may have found this useful in showing that they were genuine British citizens despite the foreign name. Harold's suited well for whaling and shipping, where the title of captain carried the impression of high marine authority.

The great expansion of the late twenties brought in an entirely new type of floating factory. Specially designed and built for their purpose, they were bigger than the converted liners and broader in the beam. They had their machine rooms aft, with the stern slip going up between twin funnels, and the bridge and crew accommodation placed well forward at the bow, leaving a large midships section for the whaling 'plan' deck where the whales were cut up, and a spacious factory area below it for the processing machinery. Below that again was large tank storage for fuel oil and whale oil. The new floating factories were more expensive than conversions, probably by 50%, but had substantial advantages. With an appropriate catcher fleet they could take and process many more whales in a season than their predecessors, at considerable savings in costs. *Kosmos* produced in its second season, 1929/30, 199,000 barrels of whale oil, while *Kosmos II* in its first season, 1932/33, provided an all-time record of 222,000 barrels. The best of Salvesen's floating factories that year, *Salvestria*, only achieved 130,000 barrels, a catch that by any other standard would have been outstanding. These new factory ships, still at the design stage when Salvesen were planning their two conversions, seemed an unjustified gamble to the Salvesen directors, who reckoned to be more careful with their own money than companies using borrowed capital. But the new design fully justified itself and became standard for whale factory ships. Our postwar ships, *Southern Venturer* and *Southern Harvester*, followed exactly the design and size of *Svend Foyn*, one of that new generation of floating factories.

Many of the companies that took part in the whaling boom were the well-established Norwegian and British operators. Lars Christensen, whose father Christen Christensen had sent the first floating factory to the Antarctic, continued to be prominent, as was Johan Rasmussen, who had sent the first whaling expeditions to the Ross Sea. To these 'whale kings' was added Anders Jahre, who skilfully promoted and shaped old and new companies. His master achievement was the floating of companies to build *Kosmos*, outstandingly successful from the beginning, and *Kosmos II*. Jahre became in the thirties and after the acknowledged leader of the Norwegian industry, as well as one of the richest men in Norway.

The fever which had gripped the Norwegian industry spread to London, where there were financiers ready to back the ambitions of the entrepreneurs across the North Sea, so a number of the new expeditions were backed by British capital, and some sailed under the British flag. This caused concern in Norway, where the press proclaimed that Norway's top industry was being sold out to speculating foreigners. But the management of the companies, even that of the two Unilever expeditions, *Southern Empress* and *Southern Princess*, remained almost entirely in Norwegian hands. Salvesen was still the only company whose management was firmly based in Britain.

Salvestria and *Sourabaya* were conventional conversions and so lacked the advantages of the purpose-built factory ships. Our two ships, being converted passenger liners, had a large centre castle, and the whales had to be cut up smaller to be drawn round the sides of the castle from the aft plan to the fore plan. Their processing capacity was as high, often higher, than on the purpose-built ships, but more cramped.

The main equipment for processing had been pressure cookers, where the bones, meat and blubber were cooked with water and steam and the oil tapped off. These were now being superseded by rotating cookers, which broke up the material under steam pressure and blew it over to settling tanks. The rotating cookers provided much more capacity for a given space than pressure cookers, and were particularly suitable for blubber and meat. The new factory ships installed large numbers of them, while keeping press cookers for bone. *Sourabaya* had in addition a meat meal plant, as did *New Sevilla*, which Salvesen acquired later. Neither of these plants was a great success. They were pioneering attempts by Norwegian firms to develop drying techniques suitable for shipboard, with steam heat. *Sourabaya* gave up production, but *New Sevilla* was given a new plant, this time by a British company, and she continued to produce meat meal till 1940. One or two Norwegians had also tried meal plants, but not for long. The price for meat meal was low and they did not think it worth the trouble and expense.

To match the new floating factories Salvesen strengthened its whalecatcher fleet. Smith's Dock of Middlesbrough delivered no less than thirteen new boats in time for the 1929/30 season. These were 12-knot ships, 116 ft. long, well ahead of the 11 knots standard in the early twenties. Harold recalled in later years that when Anders Jahre ordered whalecatchers for his new *Kosmos* he rang Smith's Dock and asked for 'the same as Salvesen'. Harold thought this a nice tribute to Salvesen's whaling expertise.

The Salvesen partners were not entirely carried away by the prevailing enthusiasm for expansion. They had longer memories, and their investment in conversions rather than the more expensive new-built factory ships showed their fear that the whaling industry might well be wiped out by overfishing within a few years. Harold, in an interview for the local Tønsberg newspaper, was pessimistic when asked about training British people for whaling. If we could rely on another twenty years, he thought, it would be worth while. But as we could hardly expect more than four or five years more of the present intensive catch, Salvesen were happy to continue with Norwegian crews. Fortunately Harold's fears of an early end to whaling, made in the wake of the huge 1929/30 catch, were not realised. Helped by the respite of the Second World War, Antarctic whaling continued in full vigour for well over the twenty years forward that he looked for, and only reached its end in the 1960s.

The 1930/31 season brought inevitable collapse. The main whale oil buyer, Unilever, could not absorb the immense increase in production, particularly in world slump conditions, and the price of whale oil halved from £25 to £12. The Norwegian companies, encouraged by Unilever, took the decision to lay up their expeditions and do no Antarctic whaling in 1931/32. Many of them could in any case not cover their costs at the low whale oil price and would have found it very difficult to raise the working capital to finance their expeditions. They urged Salvesen to join them, but Salvesen had other views. They were under no financial strain, not having been weakened, like some of their competitors, by having to sell their oil at the bottom of the market to satisfy bank creditors. Harold as a good liberal economist preached that free competition should be allowed to have its beneficial effects. The surplus of expeditions would be reduced by the bankruptcy and removal of the weakest, and those remaining could survive at low whale oil prices. On a broader front he argued, in a letter to the Colonial Secretary, 'the curtailment of the operations of one industry normally leads to corresponding curtailments in other industries, and the results for all those industries, for employment and for the exchequer are universally bad'. Harold was always concerned to set up general principles to justify his actions.

So it came about that in 1931/32 the Antarctic was clear of whalers except for Salvesen's three expeditions, *Saragossa*, *Salvestria* and *Sourabaya*, and Unilever's *Southern Empress* and *Southern Princess*. Unilever did not feel that the advice given to the Norwegians need apply to their own ships. With the whaling grounds to themselves the expeditions did very well, though *Saragossa* went on fire during the season and had to be abandoned. The chagrin of the Norwegians to see their British competitors catching whales while they were laid up at home was considerable. But Salvesen were able to make up their own minds on how to run their company, and they did not see that a restrictive practice designed to raise artificially the price of whale oil to a level that would keep the inefficient in business was any solution to the industry's problems.

There was a large increase in efficiency brought about by the drastic reduction in the number of expeditions both in the lay up year of 1931/32 and in subsequent seasons. Production from Antarctic pelagic whaling was 2,300,000 barrels in 1929/30, the year before the peak of the boom, and about the same in 1932/33, the year after collapse. In 1929/30 41 floating factories and 163 catchers produced that total; in 1932/33 it was achieved by only 17 factories and 112 catchers.

Two: 1931-1948

The lay-up season of 1931/32 proved a watershed for Antarctic whaling, bringing in a new era of consolidation and control. The whaling companies realised that they could only maintain a stable price and profitable operations if there was a large reduction in the number of whaling expeditions and a total production well below the 3.5m. barrels of the 1930/31 season. They therefore came together to make agreements to limit catch and catching effort. The basis of the agreements was the allocation of quotas to individual expeditions, with freedom to transfer and trade them between companies. That encouraged the older fleets to keep out of the game, collecting some money for their quotas instead. This quota system was very successful. The pelagic catch was stabilised at between 2.2 and 2.5m. barrels, two thirds of the peak 1930/31 figure. It did not rise again until 1937, when the arrival in force of Germany and Japan on the whaling grounds upset the balance.

To those who later blamed the greed and short-sightedness of the whaling companies for the collapse of the Antarctic whale populations it was a fair response, which I often used, that the only time in the history of the industry when there was effective control of catching was in the period of the 1930s, when it was in the hands of Norwegian and British whaling companies, without direct intervention by governments. Control was made easier by the closeness of the companies, which, despite the intense competition between them, shared the same background and interest. The Salvesen credo, eloquently preached by Harold, was that the last people to entrust with conservation of natural stocks were governments. Their interests were essentially short-term, to raise more taxes, preserve the country's balance of payments, or defend national honour. There were certainly plenty of examples of this in the history of Antarctic whaling, starting with the lax attitude of the Colonial Office towards issuing of licences in the early years and ending with the total inability of the whaling nations to agree drastic catch reductions before it was too late.

Harold Salvesen, although still only in his mid-thirties, played a leading part in the negotiations between the companies. Salvesen had strengthened their position by buying *New Sevilla* in replacement for *Saragossa*. *New Sevilla* was one of the largest of the last batch of converted floating factories, commissioned by a Norwegian company which had bought *Sevilla*. When that company's managers started to pinch Salvesen and Christensen gunners for their new fleet, the two companies cooperated to buy shares on a scale that enabled them to remove the management. Salvesen went on to buy up the whole company, which was on the verge of bankruptcy, so getting an excellent ship at a low price.

With three large and modern floating factories Salvesen were able to negotiate a decent quota for the company. While the Norwegians wanted to relate quotas to tank capacity, Harold insisted that they should be based on processing capacity, which he rightly saw as a more relevant index of efficiency. He also got the quotas expressed in whales, instead of in barrels of oil, as originally proposed. A quota in barrels would have encouraged operators to catch as many whales as possible and forget about the oil yield. A whale quota ensured that good processing and high oil yields would be rewarded in higher production. These arrangements were both rational and to Salvesen's advantage.

The Norwegian companies, represented by strong-minded successful entrepreneurs, were not easy to negotiate with. Harold found their ideas of commercial morality very different from his own. They could be discovered in the course of negotiations trying to fix conflicting private deals. With fingers in many different whaling pies, they were quite ready to agree for one company what they then repudiated for another also under their control. Harold got used to this, but ensured that the agreements were very carefully worded to block any loopholes for evasion. He clashed with Jahre, the most forceful of the Norwegian owners, who was used to getting his way and did not like being challenged by a young newcomer to the industry. Harold doubted Jahre's honesty, seeing him as a sharp financial manipulator with no real knowledge of whaling, but he had to admire Jahre's considerable business success both in whaling and in shipping.

The regulation of whaling, however, was not being left entirely to the whalers. The move to pelagic whaling had released them from the constraints of the Colonial Office régime, but growing international concern about depletion of the whale stocks, and at their wasteful processing by the floating factories, forced the governments involved to set up a new framework to control whaling in the open seas. The first to do this was naturally Norway. Its law of 1929 laid down minimum catch sizes for blue and fin whales and required efficient processing. Detailed catch statistics were to be kept, and each expedition was to carry an inspector to ensure compliance.

Britain was still trying to find ways of stopping the pelagic whalers from escaping from its coastal sovereignty, but finally it had to accept the situation and fall into line with Norway. The eventual vehicle for cooperation was the Geneva Convention of 1931, adopted through whaling laws by Norway and the U.K. In this the USA played a leading part, though their whaling interest was quite small.

The Norwegian law of 1934, which strengthened their 1929 law, had a provision enabling government to refuse to allow Norwegian whalers to take employment with foreign companies. The intention was to use the current Norwegian monopoly of whaling skills to stop new foreign expeditions entering the industry. The Antarctic already had too many whalers for the good of the whale stocks, and new entrants might well not respect the international whaling conventions. This was looked on with some suspicion by the British government, which was afraid it could be used to cripple the British whaling industry. Salvesen were less fussed about it, realising that it was not directed at them. The later attempts to exploit our dependence on Norwegian crews did not come from that law, though they were equally distasteful. In fact the 'manning paragraph' only came into its own after 1945. The German fleets that came in from 1936 onwards depended greatly on Norwegians, particularly gunners, but the Norwegian government did not feel able to offend the Nazi government by blocking its recruitment of their whalers.

From 1935 onwards both the Norwegian and British governments became actively involved in whaling control far beyond the general laws they had adopted. The Norwegian government, with its fleets still dominant in the Antarctic, felt it had the prime responsibility to regulate the catch. But there was constant pressure on it from the Norwegian companies to look after their interests. The British government was no less concerned to control catch, but also to defend its own whaling industry. Apart from Salvesen and Unilever there were several companies which whaled under the British flag, though management was almost entirely in Norway. The Salvesen aims were reasonably broad. They wanted to keep down catch, both to maintain the price of whale oil and to conserve whale stocks, and were ready to join any quota agreement which gave proper weight to the larger and more efficient expeditions and encouraged the older and smaller to retire from the industry, as they would have been forced to do in a free market. Unilever's aims were rather different. They were in the whaling business to provide raw material for their margarine factories. They saw no reason why they should restrict their catching to allow someone else to produce the oil required. As Unilever was the main buyer of oil from the Antarctic catch, it was difficult for the other whalers to put any effective pressure on them except the general warning that if no catch restrictions were agreed whalers and whale oil would quickly disappear. The Norwegian companies normally acted jointly in selling their oil to Unilever, but each was always ready to steal a march on its competitors. Lars Christensen infuriated the other Norwegian companies by contracting privately to sell a large tonnage of oil to Unilever from the 1934/35 and 1935/36 season. This cut across any possible quota agreements, and was announced shortly after he had recommended a general lay-up to improve the oil price.

Finally, the Norwegian trade unions came into the scene. In 1936, after a round of negotiations on new quotas had broken down, Anders Jahre produced a proposal for the whole fleet, including the British companies, and persuaded the Norwegian unions as well as the Norwegian government to back it. The unions had made themselves a strong position with the whaling companies and had the encouragement of a Labour government. They came to believe that unless the proposed quota agreement was accepted the whole future of Norwegian whaling, including their members' jobs, was at severe risk. They declared a blockade of any companies, Norwegian or foreign, that refused to sign up. For good measure they demanded that all whaling personnel on whaling expeditions, whether Norwegian or foreign, should be Norwegian. Wiser heads in Norway were appalled by these threats, and rightly so. There was a violent reaction in Britain, as strong from the British government as from the whaling companies. The government was not going to accept control from foreigners, whether government, industry or unions, over well-established British operations in an ocean for which Britain had always considered itself responsible. It had always wanted Salvesen to use more British crews on British ships and could hardly contemplate giving the Norwegians a monopoly there. Salvesen had always had some Britons in whaling, but from the early 1920s onwards had been content with mainly Norwegian crews. The threat of a union-applied boycott and the forced removal of their British staff was outrageous. The British government made sharp diplomatic representations. The Norwegian government tried unconvincingly to dissociate itself from the union action. The Norwegian position collapsed, the unions withdrew their boycott, and the quota proposal was scrapped.

Harold was in the middle of the fight with the Norwegian unions, and for a time became a hate figure for the press in Norway. The Norwegian seamen's union history, *Fra Trellekar Til Frie Folk*, paints him as a ruthless capitalist and attacks his father Theodore, the 'Norwegian-born Lord' (he was neither, though his brother, as a Scottish judge, was Lord Salvesen), for having exploited the Norwegian whalers on the company's ships in 1931/32 by cutting wages by 75%. In fact there was no reduction in basic pay, but bonus rates, which had been fixed at the low catch level prevailing before ice fishing developed, were reduced to a more realistic level.

The unfortunate sally into nationalism rebounded on the Norwegian unions, and both Salvesen and Unilever started to replace Norwegians in their fleet with Britons. The Norwegian proportion of Salvesen crews, which had been 95% in 1935, was down to 64% in 1938/39. Most of the recruitment of British whalers was for the factory crews. On the catchers the gunners were often allowed to choose their own mates and ABs. Strong teamwork was needed and the working language could only be Norwegian, so fewer Britons were brought in there – mainly in the engine room. Salvesen, worried about the effects on efficiency of bringing in so many inexperienced people, sent *New Sevilla* on a short and fairly unsuccessful foray to the Davies Strait in the Arctic to train their new intake. But the new arrangements settled down very well; the British whalers were mainly from Shetland and had a similar background to the Norwegians, and there was rarely anything but harmony between the two nationalities. By the time I came into whaling a 50/50 split had become an accepted norm, though Norwegians did remain the slight majority.

Harold built up good relations with the unions after the showdown of 1936. He disliked Ingvald Haugen, the leader of the

Norwegian Seamen's Union, judging him untrustworthy and spiteful, but when he started recruiting British whalers he brought in the National Union of Seamen, who represented the seamen of the British merchant navy. They were delighted to find a new field of employment for their members at rates a good deal higher than on merchant ships. Charlie Jarman, the secretary of the National Union of Seamen, became a firm friend and supporter against the nationalist claims of the Norwegian unions, to the extent that *Dagbladet*, the left-leaning and scandal newspaper of the time, accused Jarman of being in Harold Salvesen's pocket, and worse.

Harold, in collaboration with the British unions, gradually brought the British companies into the annual Norwegian pay negotiations. Salvesen contributed to and endorsed the settlements but had their own contract. Harold was particularly pleased at instigating a clause obliging the individual to do any work that was required of him. There were plenty of different skills and titles on a floating factory – flensers, lemmers, cookers, boiler emptiers, etc. – but no job demarcations, so there were never any restrictive practices on the whaling expeditions. He also removed the conflict inherent in the yearly bonus rate calculations. There had to be an assumed catch for each expedition agreed beforehand which would give the whaler his targeted bonus earnings for the season if exactly achieved. A higher catch would earn him more bonus, and there would be less if the catch fell short. The companies naturally wanted a high bonus basis, the unions low. Catch figures in previous seasons were a guide, but it seemed unfair to penalise whalers by a high basis this season just because they had been lucky last season. Harold's solution was to take an average of the last three seasons. That showed what the expedition could do over a period, smoothing the fluctuations of catch. The arrangement stuck, and was accepted without challenge for all the postwar seasons.

In the later 1930s efforts by the Norwegian and British companies and their governments to stabilise both catch and catching effort were frustrated by the arrival of the Japanese and Germans. Japan had started modern whaling from its coasts in 1905 and built up a substantial industry. Though Norway had in the early years provided people and material, the Japanese very soon became entirely self-sufficient. In 1935 they turned their attention to the Antarctic, buying an elderly Norwegian floating factory. They followed up by acquiring the plans of the latest generation of factory ships from the British builders. Harold bitterly attacked the Norwegian authorities for allowing the Japanese such an easy entry into the Antarctic, particularly as they were not ready to sign the international agreements, but the Norwegians took the view that the Japanese were determined to get into Antarctic whaling anyway and nothing could be done to stop them. By 1939 they had six floating factories, all, except the first, new and built in Japan. More were planned. This expansion was pursued with great national fervour. Japan was to out-compete Norway and Britain, and only when the Japanese had completed building up a substantial fleet would they consider accepting any international restrictions on their catch. They were certainly formidable competitors: their ships were of high specification, their crews were cheap and at least as skilled as the Norwegians, and they could also use much of the whale meat for domestic consumption. Their boast that they would dominate the Antarctic was set back by the war, but was eventually fulfilled.

The force behind German entrance into whaling was the determination of the Nazi régime to make the country self-sufficient in foodstuffs. Whale oil was the main raw material for margarine, and the German government resented having to buy its supplies from what it saw as a combine of Norwegian and British whaling companies restricting production and pushing up the price. Unilever's margarine factories were a large part of the German industry, and they came increasingly under state control. The German government, by blocking Unilever's earnings in Germany, constrained the company to build up a national whaling fleet. Unilever could hardly resist the opportunity both to release their earnings and to reduce their dependence on the independent British and Norwegian fleets. The Norwegians struggled hard to stop any expansion, trying first to apply the 1934 Whaling Law to put an embargo on the employment of Norwegians, but they were in no position to antagonise the Germans as one of their biggest whale oil buyers. No embargo was applied, and they had reluctantly to approve the purchase, by the German margarine companies, headed by Unilever, of one floating factory from their fleet, with the chartering of two more. A further purchase from England, plus two new buildings and a conversion in Germany, brought the total to nine by 1937. The two new factory ships were remarkable. One was *Walter Rau*, named after the leader of the margarine combine which owned her, the other, *Unitas*, was built by Unilever with its blocked currency. *Unitas* was the largest floating factory built to date, 30,000 tons deadweight (i.e. carrying capacity), compared with about 18,000 tons for *Svend Foyn* and her sister ships. The original design had been for *Terje Viken*, built in 1936 for the British/Norwegian company, Hector Whaling, but with an extra deck superimposed which increased her factory processing space considerably. *Walter Rau* was the first factory ship in which the best food technology was applied to use the whale's potential to the full. In addition to the normal oil producing plant *Walter Rau* had a meat meal plant, a meat freezing plant and experimental plants for blood meal, meat extract, canned meat and vitamin A oil (extracted from the liver). Not all these plants survived. Some of the products had no market, even in Germany, but this thorough and characteristically German approach to whale processing stimulated new thinking among the established operators which bore fruit in later years.

The disastrous outcome of the 1936 blockade against the British fleets had brought home to the Norwegians that their monopoly of whaling skills, established in the first years of Antarctic whaling, had a limited life. Its strength was indeed considerable. Tønsberg and Sandefjord lived and breathed whaling. It was the ambition of every bright boy to follow his family into the profession, starting as mess boy at sixteen and progressing, with studies at nautical school and certificates on the way, to become a first mate on a catcher and finally enter the hallowed ranks of the gunners. Or he might become an engine room boy and climb the ladder to chief engineer, coaxing extra knots from the catcher's steam engines and manning the winch to play the whale on the line after the gunner had called 'Fast fisk'. British whalers developed the same loyalties and ambitions over the years, though the firm dividing line between officers and ratings in the British merchant navy contrasted uneasily with the smooth path of progress in Norway. The greatest Norwegian strength lay in their gunners, on whom depended the success of the expedition. They lived by their results, and the weaker ones were replaced each season. Their skills came from the combination of a steady eye with drive, stamina and leadership. They had spent their lives on whalecatchers, many of them starting as ABs and mates to a father or uncle, eventually getting the chance of a few shots which would show whether or not they were worth promoting and trying out. The new foreign expeditions found the Norwegian gunners the most difficult to replace. The Germans did not try. They advertised for Norwegian gunners and obtained what they wanted, much to the chagrin of those in Norway who felt that Norwegians were betraying their industry by going over to foreigners who would eventually destroy it. The Japanese however took this in their stride. They started with a few Norwegian gunners, but quickly trained their own people and soon had their fleets fully manned by Japanese. Salvesen made no attempt to replace Norwegian gunners by British. Recruitment of British whalers had started after the blockade but we had never intended to go further than provide the same opportunity for British as for Norwegians to reach the top. Our gunners, some of them now following their fathers in Salvesen service, continued to be almost all Norwegian.

Throughout the thirties whaling operators had to be tough and resourceful to keep their footing in an industry which was increasingly embroiled in international politics. In the early thirties there were the consequences of the collapse of the boom which had culminated in 1930/31. Whaling costs had to come down to meet the reduced price of whale oil, surplus whaling fleets had to be forced or persuaded to get out, and there was the looming prospect of swift depletion of the whale stocks, which governments now felt must be their responsibility. The entry of Germany and Japan, determined to wrest a share of what there was from the old whaling countries, produced a new set of problems. By 1938 the catch of whales in the Antarctic had in numbers gone right above the record year of 1930/31, to a peak of 46,000 whales. The oil production of 1937/38 from that catch was 3,300,000 barrels, somewhat below the 1930/31 level. The factories were extracting more oil per whale, but as blue whales became scarcer catch had moved over to the smaller fin whales. A new international convention in 1937, accepted by Germany but not Japan, did provide some protection, but not much. Minimum sizes were introduced for catch of blue, fin and sperm whales, and most of the warm water areas of the world, as well as the Arctic, were closed to fishing of baleen whales by floating factories.

The shutting off of those areas to whaling was a sensible conservation measure. They held largely the same whales which trekked back into the Antarctic for summer feeding. While in the warmer waters they lost most of their blubber thickness and the yield of oil was very slim. The coasts of South America, Africa and Australia all provided fields where floating factories could make large catches of thin whales to little commercial benefit and at the expense of the permanent land stations. Salvesen sent *Salvestria* and *Sourabaya* to the Peruvian fishery in 1936 and 1937 on the way to the Antarctic. Some blue and fin were caught in 1936, before the new agreement had been made. But the main catch was sperm whales which were there in abundance and, except for minimum size lengths, were not controlled by international agreement.

The whalers enjoyed their Peru trips. The whales were easy to catch – one gunner shot twenty-four in eight hours – and to work up whales in shirt sleeves under the calm sunshine of the Pacific was a picnic compared with the Antarctic.

A new provision introduced in 1938 was to make a whale sanctuary, west of the Antarctic peninsula, in which whaling was banned. This was an area which the floating factories had largely left alone, as there seemed to be very few whales there. So it was not difficult to get international agreement on a conservation measure which looked important but in fact had no impact on the whalers.

These arrangements did very little to reduce the Antarctic catch. In 1938/39 this showed the first classic sign of overfishing, higher catching capacity but a lower catch. The outbreak of war in September 1939 gave a little relief, keeping the Germans, though not British, Norwegian or Japanese, at home. But after 1940/41 there was virtually no whaling until the end of the war, so the whales had a respite which allowed some stock recovery to take place.

With a fluctuating whale oil price, at its lowest a third of what it had been in the late twenties (£10 against £30), there was continued pressure on the companies in the 1930s to reduce their costs. This was achieved mainly by fewer and larger floating factory expeditions, each taking a much higher catch. Catchers became bigger and more powerful, and factories improved their oil yield. But no technical advances could take away the key importance of the gunners and of the expedition manager, who was either the master of the floating factory or the leading gunner. The manager had to be able to find whales and put his factory and catchers among them. His decisions came from his own long experience and the wisdom of his gunners, as well as from all the information which went round among the whaling community. Though companies tried to keep back news of good fishing grounds as they opened up there were no secrets between the managers and gunners during the summer in Tønsberg and Sandefjord. The local newspapers, *Tønsbergs Blad* and *Sandefjords Blad*, carried far more news and gossip about whaling than about Hitler. Some managers were conspicuously successful. Lars Andersen, who was a pioneer of ice fishing with Unilever, led *Kosmos II* in its record-breaking season of 1932/33 and was later attracted over to a German company for enormous earnings. He was bitterly attacked in Norway for betraying his country. His involvement with Germany continued during the war, so he was not taken back into the Norwegian industry after it, but as a true soldier of fortune he offered his lance to the Greek shipowner Onassis, whose *Olympic Challenger* made a brief piratical incursion into postwar whaling. Eventually he did reappear as a Norwegian manager, just before the end, but by that time his magic had gone.

Like violinists and footballers, there were plenty of competent gunners to be found but few stars. In the early years of whaling there was intense competition for the best gunners, which pushed up pay and bonus rates to very high levels. Companies were even ready to build a new catcher specially to secure a top gunner. This was damped down in 1926 when the whaling companies all agreed not to poach each other's gunners, an arrangement which held firm for forty years, up to the end of whaling. Salvesen then, like their competitors, concentrated on weeding out the duds and bringing forward new hopefuls. Most of this was in the hands of Leganger Hansen, in Leith Harbour or Tønsberg, with the help of gunner leaders such as Gustav Davidsen and Hans Abrahamsen.

Salvesen fared quite well during the thirties. The South Georgia Company record books show three years of profits in the middle years of between £200,000 and £400,000, with small profits and losses in the other years, including the 1931/32 lay-up season. The cost of the new factory ships *Salvestria* and *Sourabaya*, together with thirteen catchers, appears in the records at about £700,000. *New Sevilla* with four catchers was originally commissioned for about £300,000, but as Salvesen bought the owning company shares at a large discount the cost to them will have been substantially less. Thus the total investment, including Leith Harbour station and catchers and the whaling transports cannot have been more than, say, £1.2m. An average profit of £133,000 a year over the nine years from from 1929 to 1938 – 11% return on the original capital, rising as it was depreciated – is not exciting, but the figures are probably understated. Some of the profits each year went to the managing partnership, and it was the policy of the company to reduce its declared profit, and so its tax liability, as much as it legally could by charging depreciation of its ships at the highest possible rate. Though there were fluctuations in catch the main element in the ups and downs of profit was the price of whale oil, which swung between £10 and £14 in 1931/33 and rose to £24 in 1937. Unilever was virtually a monopoly buyer in the early thirties. That only changed when the German government came in with its own buying programme.

With the development of ice fishing a new steady pattern established itself for the Salvesen whaling operations. The nerve centre in the south was Leith Harbour. The station had continued to get good catches through the twenties and early thirties, with a peak of 101,000 barrels in 1924/25, but the strains on the whale stocks increasingly showed around South Georgia. Blue whales became scarce, and catch concentrated on fin whales and even went over to the sei whales, previously ignored as too small to be worth catching, so that in the late thirties Leith Harbour was only producing about 40,000 barrels, quite overshadowed by the three floating factory expeditions. It did, however, provide a valuable base for the whaling fleet. The floating factories called there at the beginning and end of the season, mooring side by side at Coronda Quay with their bows pushed in to the shore. Crews were exchanged, stores picked up, whale oil and fuel transferred. Leith Harbour took in most of the whalecatchers at the end of the season and repaired them during the winter in preparation for the following season. There was a slip for hauling up the smaller catchers and a dry dock at Stromness, in the next bay, for the bigger boats. The station had all the necessary shops and skills for a marine repair yard – welders, platers, riveters, carpenters, blacksmiths – and the catchers were turned out at the end of the winter in first class trim. Apart from normal overhaul and replacement of worn parts there was always operational damage to be put right. Hull plates, propellers and driving shafts got bent or bashed by ice, whales, and collisions.

The crews for the winter were recruited from the station and the floating factories. They had an eighteen-month stint in the south. Leaving Europe in the late summer they served through a whaling season, came ashore at South Georgia for the winter, and returned home in May of the following year after a second whaling season. There seemed to be no shortage of volunteers for overwintering. Winter in South Georgia was colder than summer but brighter and less windy. There were occasional heavy snowfalls, up to ten feet deep, which made a lot of work in clearing the decks of the boats, but provided good skiing.

The floating factories left Leith Harbour for home towards the end of March. By early May, having dropped their cargoes of whale oil in Rotterdam or the Mersey, they were back in Tønsberg, greeted as they tied up by their families crowding the quay. The factory ships then moved to their moorings in nearby Husøyflaket to be overhauled and painted, an attractive summer occupation for their crews. The transport ships that took the Leith Harbour crews to and from South Georgia, and the tankers serving the fleet, were mainly manned by British merchant navy crews and so repaired in the UK.

The floating factories had heavy alteration programmes each summer as new equipment was introduced and more processing plant pressed into the factory space. New catchers were ordered from Smith's Dock, increasing in size. The last prewar order, of the *Sondra* class, was for boats of 148 feet, with watertube boilers and a speed of fourteen knots.

The offices in Leith and Tønsberg were kept busy winding up the old season and preparing for the new. Whale oil production had to be sold and delivered and stores and equipment ordered. The total whaling crews at the time were about 1,400, 200 on each of the floating factories, 500 at Leith Harbour, and 14 on each of the 25 whalecatchers. The new crew lists were started on early in the summer to find who was coming back and who had to be replaced. In Tønsberg this was led by our Norwegian representative, Onésimus Andersen, and by Leganger Hansen, as much at home among the whaling entrepreneurs as in his chiefdom at Leith Harbour. In Leith the whalers called in to stake their claims at the crew department at 29 Bernard Street, conveniently placed for the King's Wark pub, locally known as the Jungle, and Mrs Moss's Chain Pier bar.

Leganger was a man of authority, whose firmness and rollicking humour could deal with the roughest of whalers. He was also generous, and had an eye for deserving cases among the many who applied to Onésimus's office for jobs in the Antarctic. One such was Trygve Bratteli, from a needy family of eleven children, who started in the guano shed at Leith Harbour and ended up as Prime Minister of Norway. Leganger was also an active and successful investor in whaling and shipping companies, riding out the boom and slump of the early 1930s. When he died fifty years ago he left a handsome sum to form a charity to help retired seamen and their dependents. Leganger Hansen's Legat has flourished and multiplied over the years, and now provides superb centres for old people, as well as sheltered housing, in Nøtterøy, his home community outside Tønsberg.

At company head office in Leith, while Noel dealt with the whale oil sales and general management, Harold was more than busy running the operational side. He had frequent trips to make over the North Sea for quota discussions, often rough, with the Norwegian companies, and for union negotiations. He continued to be closely interested in the key appointments of managers, gunners and engineers. On the technical side he required the standard of the factories and catchers to be the best in the Antarctic. He drove our superintendents, people of great competence but used to the slower time scale of merchant ships, fiercely until he got what he wanted.

The outbreak of war in September 1939 did not stop the expeditions from going south. When they got back to Europe in the spring of 1940 the Germans were in Norway, and the ships signed off their crews in British ports. The Norwegians among them were destined not to see Tønsberg for another five years. Many went off for the rest of the war into merchant ships or to forestry work in Scotland. Three British expeditions set out for the 1940/41 season, *New Sevilla*, Unilever's *Southern Empress*, and *Svend Foyn*, all under Salvesen's management. *Salvestria* had already been sunk that summer by a magnetic mine in the Forth estuary, almost within sight of the Leith office. *New Sevilla* was torpedoed shortly after leaving Liverpool, so there was no whaling for Salvesen-owned ships. The two British expeditions were joined by three Norwegians and six Japanese, a total of eleven. During the season the Germans brought off a remarkable coup. The armed cruiser *Pelikan* rounded up in the Antarctic three Norwegian floating factories, two operating and one being used as a tanker. They were brought safely back all the way to Bordeaux with eleven catchers and 130,000 barrels of whale oil, a brilliant achievement.

After 1940/41 whaling was virtually suspended for the duration of the war. Harold pleaded the importance of whale oil in food supplies, but the government decided that ships could be better used elsewhere. The factories, as well as being large oil carriers, had valuable deck space for tanks and vehicles, and the catchers were useful for patrol work and minesweeping. In any case the Admiralty could not provide the naval escorts required for the Antarctic.

Leith Harbour catchers were sailed to South Africa for war service, and the manager, Ragnvald Hansen, was left to sit the war out at South Georgia with a small crew. It was a long vigil but they were not entirely isolated, as the Argentinian station Grytviken continued to whale. Leith Harbour was defended by a 4 inch gun, sent down by the War Office at the end of 1939, which was mounted on a headland close to the station. A Lewis gun and rifles were also provided, and a defence force recruited from among the whalers. The prospects of an enemy invasion were taken seriously. Salvesen suggested that harpoon guns loaded with nails, rivets and scrap might be used, but wondered whether this would conflict with international convention. The War Office replied that there appeared to be 'some slight doubt' about the legality of using harpoon guns loaded in that way and advised that they should not be used as offensive or defensive weapons. Fortunately none of these preparations were put to the test. No German raider appeared and the Leith Harbour men kept their watch undisturbed until the arrival of Salvesen's *Saluta* in the autumn of 1945 to herald the renewal of whaling.

With whaling in abeyance, the factory ships and transports came, together with the merchant navy fleet, under the control of the Ministry of War Transport. Salvesen continued to look after them, and all the other ex-whaling ships under British flag, including Unilever's *Southern Empress* and *Southern Princess*, were passed into the company's management. Unilever then decided to sell its fleet, which Hector Whaling and Salvesen were both ready to buy. Rather than quarrel about it Harold tossed a coin with Rupert Trouton of Hector, and won.

With a tiny staff of assistants and marine superintendents in Bernard Street, Harold worked himself to the limit. The war effort depended vitally on the constant supplies of armaments, oil and food flowing across the Atlantic. With mounting losses from U-boats and mines ships were scarce, and they had to be kept in active service at all costs. Crews and stores had to be arranged, repairs supervised, times in port for loading and discharging cut to a minimum, and sailing programmes worked out against the shipping plans of the Ministry, frequently changed at short notice to meet the unpredictable emergencies of war. The floating factories were conspicuous targets for U-boats on account of their size, with their extra decks raising the hulls high above the water line. Every one of the Salvesen and Unilever floating factories was sunk, together with *Svend Foyn* and *Tafelberg*, two factories under management. With them went some of the finest officers and men, British and Norwegian, of the whaling fleets. Some of them survived the sinking of one factory only to be lost in a second. The quiet heroism of these men, who sailed cheerfully in the face of death, voyage after voyage, has been recorded elsewhere and should never be forgotten.

Harold felt the strain of this as much as if he had been himself on the bridge of floating factories in the Atlantic convoys. His senior Norwegian assistant, a rock of support, collapsed under the pressure and was tempted into fraud. He had to leave in disgrace, but Harold could not bring himself to abandon him and later brought him back into whaling at South Georgia, a haunting ghost from the years of war.

As the end of the war came in sight the company started to look at the future of its whaling. Harold of course was determined that it should go on, although there were none of the floating factories left and many of the catchers had been sunk. He persuaded the Ministries that the restoration of the whaling fleet was a high priority, particularly with a world shortage of oils and fats. The Norwegian owners were no less interested in their future. They had fared somewhat better than the British, since five of the pre-war floating factories had survived. The British floating factories had been assigned to the trans-Atlantic convoys, where the need, and the danger of sinking, was greatest. The Norwegians, through Notraship, the official organisation controlling Norwegian flag shipping, were able to resist the Ministry of War Transport's demands to put all their ships into full war service on the Atlantic run. Harold felt that they were avoiding their responsibilities. They should, like the British, have been prepared to sacrifice their ships to win the war. The Norwegians suspected that the Ministry's requests were part of a deep plot to weaken Norway's position in postwar whaling. Their government was clear that the world fleet would have to be considerably smaller than prewar, and, with the defeat of Japan and Germany, might again be divided between Britain and Norway. Its concern was that the Norwegians, for whom whaling was much more important than for Britain, should not lose their pre-eminence in the industry.

Eventually the Ministry of War Transport allotted two of its valuable building berths to new floating factories, one British and one Norwegian. The timing turned out perfectly, and the ships, contracted in mid 1944, were delivered in time for the 1945/46 whaling season. One of them was *Southern Venturer*, built for Salvesen, who had now dropped the old *S-----a* name pattern and taken on Unilever's title.

The new ships were built on the *Svend Foyn* drawings. They were smaller than the *Kosmos* ships, but the Ministry was unwilling to authorise anything bigger, and they served their purpose well. They were somewhat handicapped by having, rather than diesel engines, reciprocating steam engines to drive them, of rather low horsepower. These were of a standard type being produced for tankers and nothing better was available. Harold Salvesen and Rupert Trouton, advising on the design, were content with steam, though they would have preferred more of it. They liked the large boiler plant of a steamship, which could double in providing ample capacity for the processing factory. They were also worried about being able to run diesels with the reliability needed in the Antarctic. The Norwegian advisers took the opposite view, having had good experience of diesels from their merchant ships and seeing no problems in operating them. They realised that a separate boiler system for the processing factory was more expensive, but did not foresee as large a steam need as the British companies with their interest in high processing capacity and by-products.

Southern Venturer and the Norwegian sister ship *Norhval* came out as steamships because of the limitations of wartime production. Thereafter the Norwegians and British diverged. Hector Whaling's *Balaena* and our *Southern Harvester*, delivered in the following year, were both steamships, while the Norwegian new buildings followed the diesel line. The whalecatchers however remained mainly steam-driven in both fleets until near the end.

There is no doubt that the Norwegian path was right. The perfect floating factory would have been larger, diesel-driven and served by a fleet of diesel catchers. That would have given great fuel economy, removing the need for frequent visits by tankers to the whaling grounds to deliver fuel oil and remove whale oil. The separate power unit also enabled the ship to move regardless of processing, whereas the steamship was almost immobile when all the steam was required for 'full cook'. But by the time we felt confident about operating diesels properly in the Antarctic the remaining time span was too short to justify the large investment which a switch to diesel would have required.

The building of new factory ships went on till 1948. Salvesen brought in *Southern Harvester*, a sister to *Southern Venturer*, but with slightly bigger engines. Both came from Furness Shipbuilding at Haverton Hill on Tees, which had built *Svend Foyn* fifteen years earlier. Hector Whaling (then called United Whalers) built *Balaena* at Harland and Wolff, Belfast. The three British ships all had large by-product plants: *Venturer* and *Harvester* produced whalemeat meal, liver oil and liver meal, while *Balaena* had meal, frozen meat and meat extract. The new Norwegian ships had efficient oil-processing plants, but, in line with previous history, little or no by-products. With three new Norwegian factory ships in addition to *Norhval*, and the transfer of the German *Walter Rau* to Norway, the combined British and Norwegian fleet came to twelve expeditions, a big drop from the prewar fleet. In addition, there was *Empire Victory* (later *Abraham Larsen*), ex-*Unitas*, the biggest of the German factory ships. She was auctioned by the British government and bought by a South African company. Salvesen put in a bid for her but at a low price, as they felt they had enough on their hands with the two new ships. There were two newcomers, the USSR and Holland. Russia took over as a German prize *Vikingen* and rechristened her *Slava*. The Dutch converted a tanker to be *Willem Barendsz*. Lastly, although German whaling was eliminated, Japan, under the benevolent dictatorship of General MacArthur, was, despite loud protest from Norway and the UK, allowed back with two expeditions.

The Norwegian government sensibly put a limit on its number of expeditions. When Jahre in 1948 jumped the gun by building a third ship, *Kosmos V*, to add to their replacements *Kosmos III* and *Kosmos IV* (ex-*Walter Rau*), the government refused a whaling permit and it had to operate for most of its life as a tanker. In the UK no intervention was needed. Salvesen and Hector had their three new expeditions and were not looking for more. The established whalers were naturally unfriendly to new countries coming into the Antarctic, even if they had been allies during the war, but there was little they could do about the Dutch and Russian expeditions. The Norwegian government was determined enough to apply the 'manning paragraph' and ban Norwegians from taking employment with those expeditions, but they went ahead all the same.

The roll call of floating factory expeditions in 1948/49 after the completion of the postwar building programme was eighteen, made up of ten Norwegian, three British, one South African, one Dutch, one Russian, and two Japanese, with a total of 212 whalecatchers. This compares with 34 factory ships and 281 whalecatchers in 1938/39, the last prewar year. Much of the prewar catcher fleets had by then been replaced by larger and more powerful boats, the companies all being determinedly optimistic for the future.

In the first two postwar seasons there were many operational difficulties. New people had to be recruited and trained. The catchers just released from their war service were in poor shape, and there was a shortage of capable engineers to man them. Salvesen's available transport ships were old and had not been improved by seven years on the battle front. *Sevilla*, already forty-five years old, and rejected for Atlantic convoys because of her slow speed and unreliability, had to be put into service again as a transport for South Georgia. On the way home she shed first one, then another, blade of her propeller, and was reduced to three knots. Hans Abrahamsen, who was on board, maintained that she took so long to get from Faerder (the lighthouse at the entrance to Oslofjord) to Tønsberg, about ten miles into the fjord, that he had to have a clean shirt three times.

The problems of overfishing and falling whale stocks had been postponed by the war, but not solved. Five years' suspension of whaling had clearly improved matters, but it was difficult to make any sensible calculation about its effects when so little was known about the total whale stocks. Meetings of the whaling countries were held in London in 1944 and 1945 to prepare for the resumption of whaling. They reaffirmed the previous restrictions on minimum lengths of whales, operating period for the Antarctic season, and closed areas. The important new provision was the fixing of the overall catch. It was realised that without this, and with free entry for new expeditions, the industry would continue to expand and the gross overfishing of the 1930s would be repeated. The maximum total catch was set at 16,000 blue whale units, two fin whales or six sei whales being deemed equivalent to one blue whale. In total catch this gave between 25,000 and 35,000 whales, depending on the proportions of each type of whale in the catch. That was a considerable reduction from the last prewar season, when the catch reached 46,000 whales.

In 1946, at a meeting in Washington, a new International Whaling Convention was signed, which set up an International Whaling

Commission (IWC) to control world whaling. This was a great advance on earlier agreements, being the first to establish a permanent regulatory body. The USA took a leading part in launching the Convention and the IWC. Their distinguished scientist Dr Remington Kellogg chaired the Washington meeting and as the US Commissioner continued to be the guiding inspiration of the IWC in the following years.

The USA's unreserved championship of the conservation cause gave strength to the IWC, but the American belief in free enterprise had unfortunate effects. On their insistence there was to be full competition within the overall quota, each expedition taking as much as it could until the total catch was achieved and whaling was stopped. That seemed fair, giving no company or country a more favourable position than it could achieve by its own merits. But because the total catch was limited any expansion of the fleets brought no overall economic benefit: the catching season was merely shortened, and the catch per expedition reduced. With so much surplus capacity in the industry there was strong resistance to any cut in the catch limit, however strongly recommended by scientists, which would reduce catch per expedition even further. These defects became clearer as the years went on and whale stocks declined. Eventually, in 1962, the overall quota system was abandoned, but too late to allow effective action to be taken. The war had provided a timely break; whale stocks had obtained some respite, and whaling companies and countries had had the opportunity to start again in a new atmosphere of postwar international cooperation. The industry's life was extended for twenty years, far more than the most determined optimist expected in the 1930s. But the ideal of harvesting the Antarctic whales so that they could remain a steady source of wealth for man in future generations was not achieved.

Three: 1948-1949

I came into whaling in 1948, having, by coincidence, followed much the same path as my uncle Harold Salvesen. I went from a classical education straight into the Indian Army for three and a half years' war service, then came back to do philosophy, politics and economics at Oxford. To start with I had no firm plans about a career. It might be the bar, or the diplomatic service, or politics. It was only after I got into economics that I started being attracted by business. At first I was dead against going into the family firm, with the dependence which that implied. I admired Harold, but knew him well enough to understand that he would be a tiresome boss. Nevertheless whaling seemed interesting, and I wanted to see more of the world. There was nothing exciting coming up from the university appointments board, so when Harold invited me I stepped in. My cousin Max Harper Gow had joined straight from the army two years earlier.

I was first sent to Liverpool to learn about shipping in the office of Salvesen's agents there. I had been there only two months, scarcely enough to learn the difference between a bill of lading and a charter party, when word came that I was to go to the Antarctic with Harold Salvesen as secretary and bag carrier. Bag carrier was no empty title. Harold took with him on his Antarctic visits a large battered leather trunk stuffed with files, a full travelling office. It accompanied us from ship to ship as we passed through the Antarctic and its papers were in constant use. They provided all the administrative backup for the operation of the three expeditions, *Southern Venturer, Southern Harvester* and Leith Harbour. Included there were crew lists, ship specifications, operating instructions, notes on new plant, stores, schedules, new projects, supplemented by reams of his own notes, scribbled in pencil on pages torn from loose-leaf pads. Of particular importance were the draft programmes for movement of the tankers from fuel loading ports to and between the Antarctic expeditions.

We had scarcely boarded the plane that was to take us to Rio when he pulled out a note pad and started to instruct me on the different classes of boat in the whalecatcher fleet. The post-war building programme was by then half way to completion. A number of our catchers had survived the war, and they were supplemented by the remaining boats of the Unilever fleet, but most of them were now too small and slow for Antarctic requirements. The new catchers from Smith's Dock had the same hull as the 1937 *Sondra* class, but the builders had managed to pack in two water-tube boilers instead of one, giving the ship a full 15 knots. The two most recent deliveries were *Southern Flyer* and *Southern Foster* (the latter named for the boiler makers, Foster Wheeler). I had been at their launch at Smith Dock yard at South Bank, Middlesbrough, a novelty introduced by one of the new partners and thoroughly disapproved of by Harold, who thought that ceremonial was out of place for whalecatchers. The baptism was indeed unlucky. On her maiden voyage south *Southern Flyer* put in to bunker at St Vincent in the Cape Verde Islands. On the way out the master, cheerful from a visit ashore, misread his compass bearing and instead of heading the ship out to sea ran her into the rocks in the bay, where she became a total loss.

Just at that time a number of war-built corvettes (the Flower class) had been bought for the fleet. These had taken much of their design from whalecatchers, but were considerably larger, with more powerful engines. Two of them were converted to whalecatchers by rebuilding the front portion to provide the characteristic flared bow and high gun platform. The others were used as tow boats, but several of them were later converted in a simpler way. These corvette catchers, with a 2,750 h.p. engine giving 16 knots or more, were another jump ahead in whalecatcher power.

Harold expounded not only the merits and deficiencies of each class of catcher but their more private characters. There was, for example, *Southern Gem*. She had been built in Germany for Unilever in 1937, larger and more powerful than the later Smith's Dock class. She was intended to be an exact copy of *Southern Pride*, claimed by Harold to be the ideal catcher, lost in the war. But she never performed properly and was disliked by every gunner who had her. Heavier German steel plates were thought to be the main reason. There was *Sigfra* built at Kaldnes, our local Tønsberg yard, in 1936. The design had stability problems, and *Simbra*, a sister ship, had been lost off South Georgia in 1947, turned over by a wave when doing an acute turn during a chase. Only one of the crew, in the crow's nest at the time, had survived. *Sigfra* was modified after the accident, but she had less freeboard than the other boats and had to be handled carefully.

I was of course unable to take in all this on the spot. But Harold set much of it out in pencilled diagrams and notes, useful (when I could decipher his handwriting) for when I met and boarded these catchers in the Antarctic.

One curiosity among the files was the record of a Norwegian arbitration court's award on a claim made by us for government compensation for the wartime loss of *Strombus*, an old floating factory bought by Salvesen as a transport in 1936 and kept under the Norwegian flag. Harold, in the course of quota negotiations in that year, threatened to put a stern slip on her and send her out as a floating factory if the proposed restrictive agreement between the companies did not go through. Negotiations failed, and he characteristically kept his word, at great cost. The ship was expensively converted at Palmer's yard in Jarrow, just reopened after the slump and very inefficient, and the expedition had a miserable catch. She never fished again. Not surprisingly the Norwegian court ruled that she did not qualify as a floating factory and refused compensation.

We also studied together the crew lists, blue for *Harvester*, yellow for *Venturer*, and pink for Leith Harbour, which set out the floating factory station and catcher crews. There were about 500 on each factory ship with 200 on its twelve catchers. South Georgia, with only seven catchers, needed smaller production crews, but had additional tradesmen to cover its job as a repair base, so its total was nearer 600. Harold knew most of the senior officers well. His detailed biographies, stretching back through the war years into the 1930s, were lost on me, who had so far never met any of them. But it was a way of absorbing some of the past, which for him was as important as the present.

After a leisurely three days' flight south, overnighting at Dakar and Recife, we had a few days at Rio waiting for a chartered tanker, *Athelchief*, which called in for us on the way to South Georgia. Our hotel was on one of the inner bays, long since filled in, and the water was clean enough for the pre-breakfast swim that Harold demanded, though none of the locals seemed to share his enthusiasm. Between visiting the sights we engaged in long discussions on economic subjects. I was freshly filled with a diet of pure Keynes. The confident intellectual consensus was that he had found the solution to most of the economic problems of the age, including trade cycles and unemployment. Harold came from a pre-Keynes background and saw no merits in the new dogmas. Our concepts seemed mutually incomprehensible and we made no progress in establishing common ground. Harold's methods of discussion were in any case not conducive to that. He argued to win, not to find agreement, and was only satisfied when he had demonstrated his intellectual superiority and dominated the battlefield. If I could reconstruct these debates, I would no doubt find that my Keynesian arguments were hopelessly naive. His more historic definitions of savings, investment and capital seem, forty-five years later, to have been triumphantly reinstated.

Sometimes our debates moved to politics. Harold combined his faith in entrepreneurial capitalism with sympathies for social welfare and equality of opportunity. He claimed to be a Labour voter and was proud of having had some of the prominent intellectual Labour politicians through his hands as undergraduates at Oxford – Hugh Gaitskell, Richard Crossman, Frank Pakenham. He was scathing about their current policies and opposed any sort of nationalisation. Nevertheless he was later persuaded by George Brown, when a Labour minister, to put quite a large sum of personal capital into the ill-fated Upper Clyde Shipbuilders when the government was trying to salvage the Scottish shipbuilding industry. He also accepted high personal tax as inevitable and thought it perfectly right that a large proportion of his fortune should go to the state when he died.

He remained generally contemptuous of government, whether right or left, as dishonest or incompetent. His leanings leftwards, allowing for the constant element of perverseness, probably reflected most his dislike of the pomposity and nostalgic imperialism which he still saw in the Conservative politicians of the day. His views also showed a strong liberal conscience, still active in the third generation of commercial success.

A week's sailing brought us to South Georgia. The first dramatic sight of the island, with its rocky ranges rising straight out of the sea and up to snow-covered peaks of 9,000 feet, remains vivid in my memory. As we came closer we picked out Cape Saunders with its guiding lighthouse and sailed past it into the broad reach of Stromness Bay, which was home to our twin stations of Leith Harbour and

Stromness as well as the neighbouring Norwegian station of Husvik. Stromness Bay was deep and relatively sheltered, ideal to support the comings and goings of the Salvesen Antarctic fleet. It was not free of hazard, however. Near the entrance was a shoal patch shown up on the surface by kelp. One morning in 1952 one of our whaling tankers, *Southern Collins*, having just cast off from Leith with a full cargo of whale oil for Europe, went straight on to it and was badly holed. The master had been blinded by the sun and failed to see the danger spot, which was on the charts and well known to all ships visiting the station. Harold happened to be standing on the quay at the time watching the ship depart. He was not pleased. *Collins*, having lost much of her valuable cargo, was patched up by the station and limped home.

It was a striking experience to turn into the bay and see ahead an industrial township of white wooden barracks, timber jetties, factory buildings of corrugated sheeting, and smoking chimneys tucked in along the water's edge, with the bare mountains rising directly behind them. Plumes of steam rising from the business end of the station showed that the factory was at work. Flocks of cape pigeons were busy investigating the whale carcasses floating by the shore as they waited to be drawn up on the plan. The air held the friendly smell of cooking whale, disconcerting when first met, but soon unnoticed. At Coronda pier, the main jetty, we tied up alongside *Polar Chief*, one of the transports used for oil and for taking crews to and from the island.

Polar Chief was a capsule of whaling history. She was commissioned in 1897 as *Montcalm* on the Canadian Pacific passenger run between Liverpool and Canada. During the first war she was disguised as a battleship by the Admiralty to frighten German raiders. She then became a whaling floating factory, as successively *Rey Alfonso, Anglo-Norse*, and finally *Polar Chief*. Salvesen managed her during the Second War, when her adventures included a narrow escape from becoming a total wreck on the coast of Iceland. We bought her after the war to supplement the transport fleet. Apart from the passenger accommodation she provided she was by that time little more than a floating museum. Her tanks at a number of different levels, all leaking into each other, could not be trusted to carry whale oil. As a former greyhound of the ocean she could achieve a spanking twelve knots, but only by burning so much fuel that most of her cargo would be used up by the time she reached her destination. Her last voyage was during a brief period of very high freight rates, when she actually made a profit with a cargo of fuel oil from Aruba home, leaving a trail of leaking oil behind her as she crossed the Atlantic.

The last remnants of past glory were in the dining saloon, which still had the original mahogany panelling intact. Several of the officers' cabins opened directly out into the saloon. I was accommodated there, in the old doctor's cabin. *Montcalm* had had the honour of transporting back to England from Canada the murderer Crippen and his mistress Ethel Neve. They had been traced by wireless on the sister ship *Montrose*, arrested on arrival and shipped back. There were plenty of creaks and groans in the doctor's cabin, but the ghost of Crippen did not walk.

Polar Chief despite, or perhaps because of, her age was a cosy ship. Her steward had served in British India ships and produced curry for breakfast before the bacon and eggs. Like the other transports, she was crewed as part of our merchant navy fleet and not as a whaler, so the officers were free from the enthusiasms and disappointments of the whaling expeditions. She would soon be mobilised to deliver oil to *Venturer*, working close by in the Weddell Sea. Meanwhile the captain nervously awaited a summons from Harold to account for himself.

The hub of the station's life was the manager's villa, with the office attached. The villa was a trim Norwegian house with shabby Victorian furniture. The dining room was adorned by a painting of a florid reclining nude, said to have been sent down by Theodore Salvesen because his wife would not have it in the house. In the villa there lived, with the manager, the workmaster, the doctor and any senior visitors from the outside world. The office housed the secretaries and the radio telephone which kept contact with the catchers. Four times a day they came in turn on the air to report their position and catch. Some of them would be towing back their whales to the station, and the station foreman had to be notified when to expect them so that the plan crews were ready to work them up. The seas round South Georgia are stormy even in summer, and the reports were often of the boats sheltering in one of the fjords along the coast, such as Elsehul or Bikkjebukt. Apart from delivering their whales the boats came in regularly for bunkering and to collect new harpoons, ropes and food supplies. Their needs also came over the radio telephone between report times and were noted by the station radio officer.

The secretaries lived in a fever of activity. Ships had to be cleared in and out, people transferred between the expeditions, production and catch reports prepared, instructions and queries by telegram from head office dealt with. Every ship from the north brought a precious cargo of mail to be distributed, and best use had to be made of the few opportunities that arose for letters to be carried home.

The station looked out over the open bay of Stromness to the north east. It was fairly sheltered, certainly in comparison with the south coast, which was fully exposed to westerly gales and almost devoid of harbours. But the winds could come down in fierce squalls (williewaws) from the peaks, and ships had to be very securely moored to stop them being carried away. The buildings on the station likewise had their roofs anchored down by wire ropes. At the north end, by Coronda Peak, was Jericho, its scrap heap still adorned by the Lancashire boilers of the old station. It was a marvellous museum of obsolete whaling machinery and brave experiments that had failed. One of our later doctors collected over a dozen different types of harpoon head, all tried out at different times, from that junk heap. Vertical meal driers dating back to the earliest years, whale claws, winches, old catcher propellers and lifeboats were also to be found at Jericho, a mine for future archaeologists.

Around the villa were the kitchens, messrooms, barracks and hospital. Conditions were still fairly spartan, though they improved in later years. But sanitation was by now decent. The old latrines along the wharf dropping their burdens straight into the water had gone. The whaler relieving himself no longer risked being shot into the air by a stream of icy water directed at his bare bottom by a playful sea elephant in the water below. There was a fierce steam heating system in the barracks, fed by pipe lines below ground which spewed up little geysers of steam at odd points where the joints had got slack. The main street ran between rows of workshops and stores to the boiler and generator houses, the plan, and the processing plants. Beyond them were rows of storage tanks for whale oil and fuel oil, called Little Aruba after the Dutch refinery in the Caribbean where the ships picked up fuel. Further on again were the cemetery and football field.

The main road was known as Pig Street. Pigs from the piggery were free to snuffle along it, though they found more to interest them in the piles of whale meat on the plan. A railway passed down the street with wagons propelled by whaler power to transfer stores and in particular to deliver harpoons, ropes and wires to the catcher pier where the catchers tied up.

The blacksmith's shop was among the workshops below Pig Street. It had earlier been at Jericho, but was moved in the 1920s after it had been overwhelmed by an avalanche in which the blacksmith lost his life. There was plenty of forging work for the station to keep the smiths busy. But their most important job during the season was to straighten out and repair the harpoons which were sent up from the catchers, bent and twisted after their encounters with their victims. Whaling blacksmiths were, barring accidents, a long-lived tribe. I noted on my last visit to South Georgia in 1959 that the blacksmith at Grytviken and our own smith at Leith Harbour had been whaling since 1916 and 1917 respectively.

The flensing plan sloped down to the water's edge. Whales waiting to be processed lay belly up, with white ribbing for a fin or sei whale, mottled blue if there was a blue whale in, by now a rare event. One of the station motor boats delivered the carcasses in turn to the plan where they were pulled by the tails into position for flensing and lemming (dismembering). Blubber passed to different parts of the factory to be steam cooked in pressure cookers. The backbone, head and jaws had first to be cut into manageable bits by a hefty steam-driven bone saw. The meat was more gently treated. Heavy cooking would have dissolved the protein and reduced the solid material available for drying into meal. When they had had their prescribed cooking period the liquors in the cookers were left to settle. The top layer of oil was then tapped off. After the oil had been cooked out the mass of bone, meat and blubber fibre was raked out from the bottom of the cookers on to the wagons. They were pushed from there to the bins close by which fed the meal driers. These were of the simple type standard to many industries, a long horizontal rotating tube, heated by an open flame, which passed the material through till it emerged fully dry at the other end to be sacked and stored in the guano shed.

That simplified description does less than justice to the production processes, which even then were quite sophisticated and, for anyone studying them, endlessly fascinating. We improved them every year, moving steadily towards better recovery of oil and protein and conversion into higher valued products. The basic oil processing remained the same, by static or rotating pressure cookers. But the full recovery of oil from the cooking liquors and the preservation and upgrading of the meat protein brought into play quite different technologies, which over my fifteen years of whaling experience transformed our factories.

In this early postwar period the liquors remaining after the oil had been drawn off the cookers went straight into the sea where they dispersed leaving little trace behind them. On the way they passed through a vast basin, the '*kum*', where they fermented and released some of the oil they contained to float on the top and be skimmed off. This was as primitive a method as you could find. The kum was a dirty and smelly blot on the landscape. What oil it yielded was dark and of low quality. Even Harold, attracted by anything that could improve oil recovery, found little good to say of it. Not long afterwards it became obsolete. High speed centrifuges were developed which would recover the oil from cooking liquors, leaving only a small percentage behind, and Leith Harbour was equipped with them.

Stromness station was in the next bay to Leith, twenty minutes or so away by motor boat. It had been a fully equipped whaling station, brought up to a high standard before it was closed down in 1931. It came into its own during the winter when the catchers were put one by one in the floating dock for hull survey and repairs. During the season it had a small crew which busied itself with overhaul of the reserve boats and any emergency repairs to the operating fleet. The station foreman lived in some comfort in the old manager's villa, the same that had received Shackleton when he came across the mountains in 1916. There was a feeling of tranquillity at Stromness, as often found in out-stations. Detached from the frenzied bustle of Leith, Stromness had an excellent quay and plenty of water depth for large ships to tie up. But it was much more exposed to the williewaws than Leith. This came home to us as Harold and I were walking along the pier with the Leith Harbour manager. Harold, deep in discourse, did not notice how close he was to the edge. A violent gust put him off balance and he found himself looking straight down to the water twenty feet below. Fortunately Captain Begg, the station manager, shot out an arm and just in time grabbed him back to safety.

It was an easy walk, past the gun and round the coast, from Leith to Stromness. On the way you passed a little fresh water dam, put up to serve one of the floating factories that worked from the bay before Stromness station was built. But you had to mind your step. There were bog holes into which you could sink up to the chest and find yourself unable to get out without help. There were several solitary walkers whose excursions nearly ended in disaster.

The station had two old catchers as service boats, *Sabra* and *Stina*. They piloted in the ships, towed them out from the jetty, kept contact with the magistrate's settlement at King Edward Cove, beside Grytviken, and made themselves generally useful. I managed to get a fishing trip with *Stina*. We had two fishermen on the strength, who took fish for the station in the bay. The main species was notothenia Rossii (*havtorsk*), a reasonable eating fish if you stripped off its layer of protecting fat. There were plenty of these of small size to be had in the bay and off the rocks, but *Stina* was after the larger fish found in dense shoals further out. Our fishing trip did not go as planned. As dawn broke about thirty miles out we came on a very dead sperm whale which had lost its tail and broken away from the catcher towing it some days previously. It had to be secured and taken home, so there was no fishing. Sperm whales as they become rotten develop a violent stench. That and a steady swell overwhelmed me with sea-sickness and I took no further interest.

There were two other stations operating on the island, Husvik and Grytviken. Husvik was run by a Norwegian company, Tønsbergs Hvalfangeri. It had been laid up between 1931 and 1945, but was brought back into use after the war. Grytviken was the original South Georgia station. It had worked continuously from 1904, and was still owned and operated from Argentina. The three stations competed for the whales round South Georgia, but in every other way there was friendly cooperation. Harold made a point of calling at the other stations when he was at Leith to exchange whaling gossip and pick up new ideas.

Husvik was only forty minutes away by motor boat, past Stromness and round the point that separated the two arms of the fjord. Grytviken was a two-hour trip down the coast to Cumberland Bay. It had to be by service boat or catcher, the open sea trip being too risky for a smaller boat. On these visits we started with a thorough tour of inspection, particularly of the factory to see if anything new had been put in which we hadn't got. There were lengthy discussions on the merits of different kinds of presscooking, screens and desludgers, and I was required by Harold to make abundant notes for future action. This was followed by an enormous dinner, sometimes with the traditional South Georgia dishes of shag or reindeer, then more whaling discussion and a final sail back to Leith Harbour in the dusk.

Harold always talked Norwegian when he could, fast and fluently. He had learned it first in Arendal in the south of Norway, so had a softer pronunciation than the sharp dialect of Tønsberg. To that he added his own marked accent which defied analysis. I was still only at the start of learning Norwegian and it was often tedious listening to him holding forth at length when I was ignorant both of the subject and the language. But some of both got through and I started to learn.

The other two stations were modest places compared to the expanse of Leith Harbour and Stromness. Husvik always seemed like a rest camp in comparison, with nothing but whaling and only an occasional transport to disturb their peace. The successive managers at Husvik were keen vigorous Norwegians, always working on new technical developments, better solids separators for the factory or more powerful radio for the boats. The station provided a good test bed for the Norwegian whaling companies, who set up joint projects to look at promising new production projects, and we tapped in to them. The station had been well restored and showed little sign of having been abandoned for fourteen years. There was also a nice piece of whaling history in an old catcher from 1912, *Karakatta*, drawn up on land to act as a boilerhouse.

Grytviken, with its Argentinian ownership, was somewhat out of the main stream of Norwegian and British whaling. Gunners and catcher crews were all Norwegian, but the station whalers were mixed Norwegian and Argentinian, many of Polish origin. Information and instruction from the home company in Argentina was irregular. Ancient ships turned up without warning from Buenos Aires bearing fuel oil and stores and got into trouble with the magistrate through having the wrong, or no, papers. Part of their cargo was often illicit spirits (*puro*), which could be smuggled past the magistrate and station manager to everyone's benefit.

Grytviken had its own special business in the taking of sea elephants. It had the sole licence to catch sea elephants on the island, with a limit set at 6,000 bulls, calculated to be the sustainable yearly harvest. This figure seems to have been well judged. During the twenty years or so of postwar sealing the sea elephant stocks round South Georgia remained very stable. At a yield of two barrels per animal the sealing gave a potential yield of 12,000 barrels, a useful addition to the whaling take, but it was an arduous trade. Argentina de Pesca had two or three sealing ships commanded by captains who knew every beach and inlet of the island. They anchored offshore and the sealers landed by motor boat and prams (Norwegian dinghies). The seals were driven down to the water's edge and shot. The blubber was stripped off and pulled by motor boat out to the sealing boat. Old whalecatchers were converted to this job, one of them being *Lille Karl*, built in 1884, one of the first catchers that worked for the station.

The official government colony was at King Edward Cove, close to Grytviken. It consisted of the magistrate, a policeman/handyman and a radio operator. There was a little jail among the government buildings, but the whaling companies looked after their own well enough, and the space was used to house visiting scientists and explorers. The magistrate's main work was to clear ships from outside entering and leaving the stations. He had to see that the provisions of the International Whaling Convention were kept as regards length of season, types and sizes of whales. In practice, as he could not be in three stations at once, he trusted this to the station managers, who kept the rules meticulously, if only because they risked the station losing its licence if they broke out.

Being a South Georgia magistrate must have been a lonely and boring life, even with the occasional excitement of franking first day covers for new sets of stamps. Wives, usually the only women on the island, contributed to comfort but also sometimes to discord within the little colony. There were dangers too. Barless, the magistrate before Fleuret, who welcomed us in 1948, was swept to his death by an avalanche as he was walking along the coast between the Cove and Grytviken. A later magistrate climbing up a badly secured rope ladder to board an incoming ship in the bay was dropped into the icy water and had to be very quickly fished out.

Harold was always friendly to the magistrate, though he had little use for the Falkland Islands government, as personified by the governor. The governor was to him a symbol of useless pomposity. Harold would give me a picture of the great proconsul leaving his empire of 1,500 Falklanders to visit South Georgia in a naval ship, and processing through the stations in cocked hat and gold-braided uniform, to the amazement of the whalers. To be fair to Harold that was not a universal condemnation. He became very friendly with one of the governors and went out to stay with him when he was promoted from the Falklands to the Caribbean. But his suspicion of the Falklands authorities was at one with his general suspicion of governments. These were never to be trusted. They would without conscience break their word or reverse their decisions in a way that would never be tolerated in business. Honesty and integrity, qualities which he admired, were concepts foreign to government officials, dependent as they were on their political masters. The more high-minded they seemed to be, the more suspicious you should be of them.

Among the transport ships used by Grytviken was *Ernesto Tornquist*, another ex-floating factory. In October 1950 she ran aground in a gale and snowstorm a few miles north of Leith, with many of the crews destined for Grytviken on board, as well as several of our own Leith Harbour gunners. By the utmost good fortune she struck at the only spot on the coast where it was possible to walk ashore dry shod and there were no casualties.

Grytviken was privileged above the other stations in having its own church. It was a handsome wooden building with a spire, put up by the Norwegian Seamen's Mission in 1913, and there were occasional postings of pastors from Norway to support it. But it played little part in the lives of the whalers. If there were no whales on Sundays they had hard-earned leisure, and if that was to be broken into for religious observance they wanted overtime payments.

My impressions of Leith Harbour in 1948 were of a somewhat shabby, but reasonably clean and cheerful place. Food was about as varied as it could be. There was little fresh fruit or vegetables; frozen foods, hardly developed then, came in later. Plenty of potatoes were supplied, and there was an excellent bakery. Local supplements were whale steaks, *havtorsk*, and sometimes penguin eggs. Cooking was also variable. Norwegian cooks were a good deal better than their British counterparts, since the Norwegian merchant navy required far higher standards for catering personnel than the British and that showed in their performance. (I later tried to do something about improving the standards for British merchant navy cooks, without much success.) There was always some dissatisfaction arising from the difference between Norwegian and British tastes. British whalers loathed fishballs and did not take to boiled meat; Norwegians saw no merit in steamed puddings.

Not all those who went whaling shared my fairly rosy view of

Leith Harbour. Dr R. B. Robertson, who sailed as doctor on one of the floating factories in 1949/50 and produced a highly coloured account of his experiences in *Of Whales and Men*, pulled out all the stops:

> Leith Harbour is the foulest place I have seen in a far-travelled life. I have been in Indian and African villages where sanitation is an unknown word, and where the villages raise themselves some inches higher every year on their own excreta; Leith Harbour is filthier than these. I have been in the slums of Glasgow, Cairo, Calcutta and Shanghai; the denizens of these terrifying slums are better cared for by their fellow citizens than are the whalemen living in Leith Harbour cared for by anyone. The whalemen's dwellings are burrows, stinking and sickening.

Captain William McLaughlin, who had served as a tanker master with Salvesen and as chief officer of *Southern Harvester*, wrote in his *Call to the South* that a shore-based whaling station was probably 'about the filthiest habitation of men the whole world over'. But the theme of whaling encourages fantasy, and both authors indulged in it. Robertson's book aroused considerable resentment among our whalers, leading the Norwegian Seamen's Union to publish an official protest. Leith Harbour whalers came back enthusiastically season after season, although there were plenty of jobs to be had at home or on merchant ships. Certainly in the many taped interviews with whalers which were collected after the end of whaling there were few criticisms of food or amenities.

The main recreations at Leith Harbour came from the cinema, the library and the football ground. The library was well stocked in English and Norwegian. I found there once a full set of the works of Strindberg, a bit gloomy for winter reading. The cinema got a new supply of films every season which were exchanged with the floating factories as opportunity arose. The football ground was rather stony, but flat enough for good games. There was a yearly sports meeting of the three stations and an interstation football tournament. In winter there was skiing, with a proper Norwegian ski jump. You could walk along the coast to Stromness, and further, though less easily, to Husvik in the next bay. To the north a stiff climb up to the ridge of Coronda opened up Paradise and Hercules Bays with their colonies of penguins, and a flat plain of glacial moraine beyond them. But there was not much time for leisure activities. Sunday work was paid at overtime rates, and most people were more interested in laying up earnings than in exploring the island.

Alcohol was generally forbidden on the stations and ships except for a ration at Christmas and on special occasions. The manager, the gunners and the doctor were allowed to be exceptions, each having a limited supply. This ban encouraged interest in the techniques of private distilling, which the manager tried to control by occasional barrack searches. Currants and yeast were much in demand. The bakery's yeast supplies had to be sent down in unmarked cases to avoid their disappearing before they had arrived at the steward's store. Other sources of alcohol, including hospital supplies, had to be closely guarded. Sometimes there turned up in the 'slop chest' (the company shop) brands of hair oil which enjoyed heavy sales until the reason for their popularity was discovered. In spite of the restrictions drink was a constant problem for the managers, and there were always some whalers who lost their jobs because of it. The worst was the compass drinker, fortunately rare, who drained the neat alcohol from the ships' compasses.

Harold had recurrent worries in the first postwar seasons dealing with drunk masters and gunners, some of them affected by the strains of the war. On one occasion he had to disarm a vicious and drunk gunner armed with a knife and out to murder a fellow gunner after their catchers had collided chasing the same whale. In my time things had become more settled. But on one occasion one of our senior masters reported to me at South Georgia that another tanker master, a strongly religious teetotaller, was selling his ship's stocks of drink to whalers on the station. Knowing the master concerned quite well I was not entirely surprised, and took action. I was grateful to the captain who had tipped me off. Harold used to complain that too often serious cases of drunkenness within the fleet were hushed up in sympathy with the man concerned and the company did not get to hear of them till too late.

The standard clothing at Leith Harbour was boiler suits outside thick Norwegian jerseys and tucked into leather or rubber boots. These had to be thigh length for plan workers. The slop chest had an excellent canvas 'timberman's' jacket, full of large pockets, which many of the officers wore. Harold and I both used khaki battle dress from the war, with gas capes to put on against rain and dirt. He also gave me a sealskin jacket which had protected merchant navy stern gunners. A fur hat with ear flaps was essential for working on a catcher. I found mine, together with a pair of excellent leather boots, in a war surplus outfitter in Liverpool. There was of course no sartorial formality in whaling. Our merchant navy captains on the tankers and the floating factory managers would usually wear the appropriate cap and badge on public occasions, but that was all.

Harold and I only had a few days at Leith Harbour before *Polar Chief* had to sail for *Southern Venturer*, giving us a convenient opportunity to get there. *Venturer* was fishing that season in the Weddell Sea, three or four days to the south, an area sometimes called the Salvesen grounds because it was a favourite for our floating factories. It was close to our supply lines through South Georgia, and was still providing good catches. *Southern Harvester* however was far away in the Ross Sea area, where prospects seemed particularly attractive.

We arrived at *Venturer* to find her on 'full cook', wreathed in steam plumes from the cooking, with side discharges spouting out brown whale liquors and cooked bone into the sea, and with the enveloping perfume of whale. *Polar Chief* tied up alongside and we were delivered on board the factory. It could have been a messy arrival if we had had to step out on to the main deck, covered with heaps of bone, blubber and meat in a general swill of bloody liquors. But we came on board by the standard route of a large basket, of a type previously used for coal transfer, hoisted by a derrick. You stepped in on the deck of the transport and were lifted up twenty feet or so to the level of the plan deck, then deposited on a platform of the forecastle building, above the bloody fray. By the same route there came aboard our kit bags, the battered office trunk, and the sacks of mail from home, to be taken off at all speed by the ship's secretaries for distribution.

The mating of two ships in open sea was not a simple operation. They had to be firmly moored together to allow the transfer of fuel oil and whale oil between them by flexible lines. There also had to be a cushion between the hulls of the ships to take up the different rhythms of their movement, which could cause damage if they bashed against each other. The factory, in essence a large deep-draughted box, was very stable, and hardly moved even in a strong swell, but the tankers could be much more lively. We traditionally used whales for fenders between the ships and they did the job very well, but they were expensive shock absorbers. If a whale was used for more than a day it started to rot, and oil and meal yield was lost when it was processed later. We went on with whale fenders for a long time, not being able to find anything better. Eventually, in the late 1950s, the Japanese came along with a specially designed large rubber fender, which gave the whaling fleets what they needed.

Tanker visits were a necessary part of operations, but a nuisance for the floating factory. While a tanker was alongside the factory could not move to follow the whales, and the catchers might lose valuable time having to tow whales in from long distances. The tanker service for the fleet had to be planned so that there were as few visits as possible during the season. The initial programme was carefully worked out at home in Leith before the season. It had to ensure that the factories and Leith Harbour were kept supplied with fuel oil while the whale oil they produced was shipped away in such quantity that the expeditions would never run out of tank space. The tanks on the floating factories and most of the tankers were interchangeable between fuel oil and whale oil, but they had to be painstakingly cleaned after fuel oil to prevent any contamination of the whale oil. Cleaning took time, and a margin of empty tankage had to be left to allow for that. Fuel oil consumpton on the expeditions could be predicted, but not whale oil production. Weekly catch could vary between 2,000 barrels (330 tons) and 20,000 (3,300 tons). There had to be flexibility in the programme to accommodate a run of steady high production.

Another complication was that the meat meal produced by the floating factory was stored in the tanks. Once there it was not disturbed until the ship arrived back at Liverpool at the end of the season. So as the season progressed more and more tank space was taken up by meat meal and there was less left for fuel oil and whale oil.

These programmes were set out in large graphs, which I carried, rolled up and secured by rubber bands, through the Antarctic in Harold's wake. We pored over them each week after the production and movement reports came in to see what changes were needed in dispositions. It was my job to keep the graphs up to date with a set of coloured crayons.

In the 1948/49 season operation of the tanker programme was particularly difficult. We had two relatively good ships, *Southern Opal* and *Southern Collins*, both war-built ships, steam-powered, of 16,000 tons deadweight. But they had to be supplemented by *Polar Chief*, too small and inefficient to do the job, and *Saluta*, which as

Southern King had been a crew transport for Unilever's South Georgia station at Prins Olav Harbour. She had worked for a period during the war carrying passengers across the Atlantic, not to the entire satisfaction of those who had expected a trip in the *Queen Mary*. By the end of the war she was already forty years old. *Saluta* was sent to collect whale oil from *Harvester* far out to the east near the Ross Sea and return to South Georgia. On the way back she came into a succession of fierce westerly gales and slowed to walking pace. The storms were so continuous that the master became convinced that the ship was doomed and had the whole crew on deck in their lifejackets and on their knees. Eventually, after over a month's voyage, she reached Leith Harbour safely.

Another crew transport and tanker was *Southern Garden*, dating from 1919. As *Gedania*, under the German flag, she had been given an extra deck during the war with space for stores and accommodation to enable her to service naval raiders. Salvesen bought her in 1948 and sent her for overhaul in Liverpool. True to the traditions of the port, the shipyard did a thoroughly incompetent job, crowned with a gigantic bill. *Garden* struggled down to South Georgia. On the way back the engines broke down completely and she had to be towed home from Tenerife.

Saluta and *Polar Chief* went to their last resting places in 1951 and 1952, being by then forty-five and fifty-four years old. *Garden* lasted till 1960. Apart from the difficulties of supplying the whaling fleet properly with these museum pieces, I used to feel highly embarrassed when I compared them with the gleaming new diesel tankers of our whaling competitors. I did not dare to discuss the rationale of ancient ships with Harold until some years afterwards, when *Polar Chief* and *Saluta* had long gone to their graves. They certainly had some sentimental attraction for him, and he was not as convinced as the rest of us that new supply ships would be significantly more efficient. Though the company had put out many millions after the war to give Salvesen as fine a whaling fleet as any in the Antarctic his instinct was to husband the company's capital. He would not have contemplated borrowing money to build new ships, as the Norwegians did, particularly for an industry which he judged had a very short life ahead of it. The old ships had passenger-carrying capacity to transport the Leith Harbour whalers south and north, and that had to be provided in some other way before they could be dispensed with. There was also the underlying fear that if we operated diesel tankers far from home they would break down. There was a severe shortage of diesel ships' engineers in Britain throughout the 1950s, and the fate of the diesel tankers we did have, and of the whaling expeditions which they served, was on several occasions dependent on the performance of an inadequately trained, doubtfully competent, and only intermittently sober second engineer.

We had only been on board *Venturer* for three days when the manager died suddenly of a heart attack. That was Hjalmar Nilsen, one of the Salvesen veterans. The chief officer was very nervous on being suddenly promoted to full responsibility for the expedition with its 650 souls, but in the event acquitted himself very well. The primary job of the manager, who in our organisation was also captain of the ship, was to oversee the navigation and safety of the ship and take key decisions on the movements of the factory in search of whales. In this he was guided by the leading gunner, Hans Abrahamsen for *Venturer*. It was rare that a manager disagreed with his leading gunner or failed to take his advice. Apart from that he had no direct responsibilities except of supervision. The work on the plan was looked after by the chief officer, the engine room by the chief engineer, the factory by the production engineer, general administration by the secretary, and organisation of the catchers by the navigating officers. So the death of the manager did not affect day-to-day operations any more than the sudden loss of a managing director would do in a well-run industrial company. But the manager was the leader of the expedition. There could be an enormous difference in results between active managers, who would stimulate and improve every part of the expedition's work, and the relatively passive ship's captain.

In a later season the manager of one of the expeditions had a nervous breakdown on the way south. We had to decide whether to replace him by flying out a previous manager who had something of a cloud over him, or to leave him in place in the hope that he would recover. We chose the latter, having taken advice by radio from the ship's doctor. The season turned out a good one. The manager did not in the event contribute much, but the organisation carried on.

Each floating factory at that time had a fleet of twelve or thirteen boats to supply it. Nine of those were full catchers and the rest buoy boats or tow boats. The buoy boat's first job was to hold whales which were waiting to be worked up. A few whales could be moored at the stern of the factory but any more than that had to be held separately, so a buoy boat was often to be seen close to the factory holding two or three whales on each side, moored by the tail. Buoy boats were also used to collect whales from the catchers in the fishing areas and deliver them to the factory. A buoy boat often had an aspiring gunner as skipper, who was able to try his hand when not required for service. The tow boats were ex-naval corvettes. They relieved the catchers from having to tow in all their whales at the expense of catching time. Later all the corvettes were converted to catchers and we thus had only catchers and buoy boats, with no boats restricted to towing.

The catchers were in full action, searching and chasing, killing and towing, throughout the hours of daylight. The gunner himself was almost continuously on the bridge, which was open to the weather, scanning the sea for whales, or at the gun platform at the bow, linked to the bridge by a catwalk. Some whales could be easily approached and harpooned. Others were more difficult, and he might spend long periods at the gun chasing and manoeuvring to get within fifty yards, the normal range of the harpoon. That was no picnic if there were heavy seas, with freezing spray coming over the flared bow.

The routines of catching had been developed over the years. The bridge was manned by the gunner and his two mates. Two ABs alternated in hourly stretches between the wheel and the crow's nest, and were relieved every four hours by the other two ABs. When the gun was fired everyone sprang to new action stations. The chief engineer came forward to operate the winch, which reeled in the harpoon line. The mates and deck crew reloaded the gun and coiled beside it the foregoer, the lighter nylon rope which was carried out by the harpoon and took the first strain from the whale when it was shot. If there had been a miss there was only the harpoon and foregoer to be recovered and the gun to be reloaded, all at high speed so as not to lose the opportunity of a second shot. If the whale was hit, '*fast fisk*', it was reeled in at the end of its line. Many of the shots killed the whale outright and it could be brought under the bows to be inflated with an air hose to make it float and then either moored to the side of the catcher for towing or cast adrift with a marking flag. If the whale was only wounded it exerted its considerable strength against the ship. The heavy whale line which backed the foregoers was paid out, and sudden strains were taken up by the system of pulleys on the masts connected to stout springs below deck. The chief engineer at the winch controlled the playing of the whale. The whale would eventually be drawn close to the ship and despatched by a second killer harpoon, with no rope attached.

When a whale had been caught and inflated the catcher would normally leave it floating, to be collected by a tow boat or recovered at the end of the day with the rest of the boat's catch. On a good day there could be as many as six blue or fin whales. They were marked with tall flags showing the number of the catcher and with a small light for night recovery. They also had radio beacons attached, on to which the catcher could home, and a radar reflector. But flags could be blown down and beacons knocked out by heavy seas, so it was always a worry searching for flagged whales, compounded by fog and darkness. Some inevitably got lost, or were only found days later, having lost their identity marks, as *dauhval* (dead whales). Sometimes catchers came on the flagged whales of competing expeditions working in the same area. These were always respected, following a whaler's code of honour that stretched back into previous centuries. But a lost *dauhval* was finder's property.

The introduction of radar after the war reduced many of the hazards of navigation and catching in the Antarctic. The catchers already had direction finders by which they could take a bearing on the factory's radio beacon and steer on it. They could now see the ship up to ten miles away on their radar screens. Most icebergs could be picked up on radar, and flagged whales would show up in calm weather through their radar reflectors. So fog, which previously could bring operations to a halt, no longer had the same terrors. Radar was particularly valuable for the catchers working out of Leith Harbour. The jagged coastline outlined on the screen showed the gunners exactly where they were and allowed them to keep well away from land. The first radars on the catchers were wartime naval sets, which were quickly superseded by improved models with higher sensitivity and larger screens.

The gunner had to have all the skill, courage, and stamina of the harpooners of past eras. He stood behind a weighty gun on a heaving gun platform, swinging it to aim along crude sights at a whale whose back broke the surface for only a few seconds before it disappeared again. Some of the gunners were fine shots, and could hit at a hundred yards. Others relied on their skills in getting the catcher placed so that the whale surfaced close under the bow. All catcher people quickly learned to spot whale blows in the grey Antarctic seascape and to distinguish between the different shaped blows of blues, fins, humpbacks, sei and sperm whales. They had to recognise and ignore the blow of the minke whale, quite abundant even in the time of the big whales but too small to be of interest to the whaler. The experienced gunner acquired a feel for each sort of whale, its social habits, where it might be found in relation to the swarms of krill on which it fed, how it would run or sound when chased, where it could be expected to surface. This knowledge, passed down through generations, was the essential fieldcraft of whaling. To it had to be added determination, endurance and patience. A few star gunners stood out from the rest season after season with talents that were much more than the sum of those qualities. On them depended the whole success of the venture.

As government regulations came in to control both the type and the size of whale allowed to be killed, gunners had to acquire new skills. From 1938 onwards there were minimum size limits for catching all types of whale. These were strictly enforced by a government whaling inspector on the factory ship, and no bonus was paid on undersized whales. When these limits were first introduced there was great concern among whalers on how they could possibly be observed. A cartoon at the time in one of the local Norwegian papers showed a whaler hanging at the end of a long fishing rod held by the gunner over the bow of his catcher. The whaler carries a measuring tape which he has put along the back of a whale that has just surfaced, and the gunner shouts to him 'Is he big enough?' In fact gunners learned quickly, from the look of the whale and the broadness of its back, to spot if it was undersized and leave it alone. Though there had been stories prewar of government inspectors being locked into their cabins for the season with a crate of whisky, there was no doubt about the integrity of postwar inspectors on the British and Norwegian expeditions. Occasionally the measuring tape on the factory plan deck might have been a little elastic.

The catcher crews were in a different world from the people on the floating factory. The factory was a vast butchery-cum-processing plant which happened to be floating in the Antarctic. You could live and work on board, absorbed with the daily routine of taking in whales and reducing them to oil and meal, without noticing the sea at all. The boiler emptiers, first flensers, bonesawmen, production officers, had routines little different from what they would have been had the whole installation been set up in their own towns, though the journey to work was a good deal shorter. The catcher men were true seamen, battling with the forces of nature, ice, fog, storms and cold, to wrest from her the prizes she could yield. Each crew was a self-contained society, happy to be left to get on with its work. The catchers had to return to the factory ship every two or three days to replenish bunkers and stores or deliver whales. It was a penance for them to tie up alongside the factory with the plan deck twenty feet above them, an inferno of dirt, smoke and smell, with orifices spouting out hot cooking liquor or bone grax on to their clean decks. Like the country dweller after a day in the city they got away as soon as they could, back to clear fresh air, birds and whales. The only time when the factory found favour with the catcher men was when at the end of the season it took them on board at Leith Harbour for the journey home. By that time the blood- and oil-soaked planking that had sheathed the plan during the season had been stripped and thrown overboard, and the ship was sweet and clean, ready for its leisurely voyage north through the tropical sunshine.

When fishing was good and the factory was at 'full cook' the ship was taking up whales for processing at the rate of forty fins a day, one every half hour. The whale was drawn up the slip by a powerful winch placed on the central bridge, 'Hell's gates', which divided the fore and aft plans. The cable from the winch was fastened to the whale claw, which clasped the whale's tail like a giant pincer. The whale was flensed of its blubber and its back meat on the aft plan, then passed through the central bridge to the larger fore plan, where the rest of the carcass was dismembered and cut into small enough pieces to go down through round holes to the cookers below. So at full cook there was always a whale being flensed on the after plan and its predecessor being cut up on the fore plan. The whole plan was a perpetual morass of blubber strips, blood, meat, bone and guts (the latter were swept over the side, sometimes disgorging a stomachful of krill). Material was moved to the cookers by hooks and wires operated by electric capstans at the edge of the plan. The plan workers had always to keep a wary eye so as not to tread on a chunk of whale as it was pulled to the mouth of a cooker, or they might find themselves down the hole. But they were trained to work safely with the wires, and the spiked soles of their long leather boots allowed them to tread easily over this treacherous and slippery landscape. For everyone else it was impassable. Those wanting to move from the messes forward to the engine room department aft had to go below and thread their way through a jumble of pipes, tanks and machines in the atmosphere of Calcutta before the monsoon.

Harold spent long hours with me on the factory deck studying the plant at work, with sweat dripping down his face. The basic oil-cooking equipment was standard, and there was not much to be seen on the outside of the cookers and the pipes that linked them to show the tumultuous conversions taking place within. The cooking process created large quantities of 'glue water', a brown mixture of water and protein with some oil. The protein was too dilute to be recovered economically, but there were centrifugal machines being developed to win back the oil, improving greatly the overall yield per whale. These claimed our close attention. There was also the meat meal plant, a chain of processes to cook the meat, press out its oil, dry and sack it. Crammed into the fore end of the factory, in a labyrinthine jumble of cookers and conveyors, the plant offered infinite scope for improvement. There was one man on the meat meal plant whose job was to stand above a hopper leading to the presses and ram down the meat into the press with a wooden paddle. I got into conversation with him and asked him why he decided to go whaling. I, naively, expected him to tell me how much he enjoyed the interest and excitement of whaling. I was quite put out when he told me he only did it for the money. He added that it was the most boring job he had ever done. So much for my idealism. But at least I discovered, as he no doubt had long before, that the job was unworthy of human labour and could be mechanised; and mechanise it we did.

Our policy in improving the production equipment was shaped by Harold's definition of waste. Waste, he declared, was an economic, not a moral, concept. There was no waste in discarding material which could not be economically recovered. That robust definition was the centrepoint of an address he gave to the Scottish Society of Arts in about 1953, intended to combat loose thinking about wasteful whaling methods. The stripping of the blubber alone by the early Norwegian expeditions, leaving the carcasses to rot, offended Harold, as it had his father, only because more oil could well have been recovered by working up the meat and bone. He felt no conscience about the tonnages of rich soup and cooked out bones which daily went over the side from the floating factory, nor was he touched by the since fashionable worries about environmental pollution. But his practice went further than his principles. His passion for efficiency sometimes led us to recover material for products which the cool judgement of the slide rule would have rejected.

The most satisfying part of the factory was the separator room. There a row of gleaming centrifuges separated the remaining traces of water and protein from the oil. The purified oil gushed out from them to be pumped to the tanks below. The capacity of the factory at full cook was 3,000 barrels a day (500 tons), so twenty tons an hour were passing through the machines. Those golden torrents of oil were the end of the complex production line, human and mechanical, the goal of the enterprise. It calmed and reassured the business soul to see success in such concrete form.

This precious product had to be stored in the ship's hull, divided like a normal tanker into large tank spaces, each holding about 500 tons. The oil might be transferred from there to one of the supply tankers which would carry it home, usually to the Unilever margarine factory at Bromborough up the Mersey. It might be kept on board till the end of the season and carried home by the factory ship itself. The tanks alternated between fuel oil and whale oil, as required. When a tank emptied of fuel oil was reserved for whale oil it had to be very carefully cleaned – whale oil even in crude form demanded the purity of a food material. A tiny contamination of fuel oil could affect the quality of the whale oil and bring massive claims from the buyers. Tank cleaning was initially by steam and caustic soda; later a machine was developed with rotating high-pressure jets of hot water. It scoured the tank walls very well, but it could not get under the struts and stringers and frames which were necessary parts of the ship's construction, nor could it clean the heating coils in the bottom needed to warm the oil into a fluid state for pumping. That task was in the hands of the tank gang, which chipped, scraped and mopped away all the bits of fuel oil which the machines could not reach. It was an arduous, though, surprisingly, not particularly unpopular job, perhaps because it paid good overtime. The tank whalers spent their lives moving from tank to tank, in hot and stuffy conditions, climbing up and down thirty-foot access ladders and perching on ledges half way up the sides so that the ship's geometry could be thoroughly covered. The cleaned tank was inspected by the chief officer and passed for use.

There was a similar cleaning job in the engine room. The ship depended on a massive daily production of fresh water for the factory processes. That was provided by a battery of evaporators which distilled fresh water from sea water. The evaporators were heated by steam coils. These quite quickly lost their effectiveness by getting encrusted with salt, so there was a gang of firemen whose sole job it was to remove the heating coils from the evaporators at intervals and bash off the hard coatings of salt. The rough treatment needed battered the coils out of shape and shortened their working lives. Fortunately a few years after the war a chemical was developed which kept the salt in solution, so the coils no longer needed hand scaling and the evaporator gang could be put to more rewarding work.

Four: 1949–1954

The 1948/49 season had seen the launching of a new product from the Salvesen fleets – frozen whale meat. Meat was severely rationed in postwar Britain, and the Minister of Food, John Strachey, appealed for help to the whaling companies. The idea seemed attractive. Fresh whale meat has all the qualities of the best beef, though it needs a little care in preparation to get rid of the fishy taste carried by the oil it holds. A good fin whale can give up to ten tons of back meat. So Christian Salvesen and Hector Whaling each bought a refrigerated ship and fitted it out to freeze whale meat in the Antarctic, all in a great hurry. Suitable ships were not easy to come by. The best we could procure was one built in the USA in 1918, allegedly to carry back the corpses of American soldiers killed in the war. She had been laid up for most of the intervening thirty years, with a short period of service during the Second World War. *Southern Raven*, as she was named, was operated by an antiquated and feeble steam turbine, and her refrigerating machinery had museum status. She had to be fitted with extra plant to freeze the fresh meat and with new compressors to operate it. Included in the diverse freezing plant, some of it unproven, was the K plant. The K plant was simply a very cold room with trays. It was put in on Harold's insistence that there should be at least one unit that could be guaranteed to work. He christened it the '*kutcha*', or K plant, *kutcha* being in Hindi the opposite of *pukkha*. In the event the K plant produced far more than any of the others.

Raven, like *Garden*, suffered a nightmarish Liverpool conversion, crowned by breaking down and going aground on a sandbank during the trial trip in the Mersey. Working with *Southern Venturer* on the whaling grounds she was supplied by one of the smaller catchers, *Sabra*, converted to ferry meat from the factory. That first season almost everything went wrong. There were extreme frustrations in using a meat transfer system that had to be painfully worked out on the job, new and untried freezing plant, and a power unit in the ship that was continually breaking down. The season's production, a little over 1,000 tons, was a long way below expectations, and a fraction of the potential material available. Harold plunged into the production systems and helped to bring some order into them, but the defects of the ship and equipment were beyond remedy on the spot. Someone had to carry the can for the mess, and that someone was the superintendent engineer in Leith responsible for the conversion. Harold on his return home discharged his wrath on the luckless superintendent, probably quite unfairly, and he had to go.

Whale meat was never accepted by the British public. However short of meat they might be they were not going to eat meat that tasted of fish. Turning it into sausages or canning it was no better. So the cargo of frozen meat from *Venturer* could only be sold, at a far lower price, for pet food. *Raven* did operate a second season, and achieved a better production, but an operation for dogs and cats could not cover costs. She was withdrawn and eventually, in 1952, sold to the breakers.

Hector Whaling's *Balaena* did have more success with producing frozen meat for pet food and continued for several years. Her freezing ship, *Bransfield*, was a good deal better than *Raven*, and there was freezing capacity on the factory itself. But frozen meat never became an important product in the European whaling fleets. It was only the Japanese who were able to develop its full potential.

When Harold, with myself trailing in his wake, was not visiting the catchers or studying the factory there was plenty of work to be done elsewhere. He was in constant touch with head office in Leith, from which came a steady stream of telegrams – movements of the fleet, queries on crew, discussion on whale oil sales, information about the catch of competitors. The radio officers, on watch round the clock, were the centre of our web of communication with the Antarctic and the rest of the world. I used to sit in their station behind the bridge, watching them taking down the morse messages and wondering what new excitement the day's batch would provide. Sometimes the messages were too confidential to be sent in clear. The code was a complicated letter jumbling devised by head office. Deciphering was my job. Occasionally a group was lost in transmission, and I cracked my head trying to find out why it wouldn't disentangle.

The competing whaling expeditions were very secretive about their catch. Any expedition that made public its success would find others in the vicinity crowding on to its fishing grounds. Figures of catch were released in Tønsberg and Sandefjord only after a lag of several weeks. But there were quicker ways of finding out about a competitor's catch. One was through the Antarctic airwaves, alive with radio contacts between factory ships and their catcher fleets, and by interception (illegal) of morse messages between fleets and home bases. Sometimes catchers of competitors turned up at Leith Harbour to be docked for a change of propeller after an accident in the ice. That gave another opportunity for exchanging information.

I recollect during the 1948/49 season several days of thick fog, not uncommon in the Weddell Sea. On one such day Harold and I took a catcher to visit *Balaena*, operating perhaps thirty miles away. Harold had picked up the habit of calling on competitors in his early seasons in the Antarctic. He was not quite so welcome in the more intense competition of the postwar era, but he invited himself all the same. The manager, Per Virik, one of the stars of whaling, received us warmly, sent us round the factory, with myself in the background scribbling notes, and gave us a good lunch with an impressive team of technical and radio experts.

When the fog swirled over the deserted plan of *Venturer* and the factory was silent, Harold would sit in the upper officers' mess, hold the table with a glittering eye and discourse on whaling, past, present and future. Breakfast was cleared away, morning coffee came and went, midday dinner was set, as he talked on, fed by tentative questions from his audience. Those who dared to comment or argue were floored by robust contradiction. Against his enormous knowledge of whaling in all its aspects there was no contest. He was particularly rough with the experts, the chemist, or production engineer, or meat inspector, even the doctor. Very few of them passed his standards. He would harass them without mercy to show up the ignorance, wrong-headedness and stupidity of people in their trades and, by implication, themselves. He was kinder to the ships' officers. They shared with him the fellowship of generations of whaling, and were still tinged with the heroic colours of the war. Conversations with them flowed back into history, reminiscences of achievements in the Antarctic and gallantry on the Atlantic convoys. I of course attended all these sessions. I used to emerge flattened by his weight of intellectual assertion, though I drank in the knowledge that carried it. At that level his passion for truth drove him beyond the normal channels of human sympathy. But at another level he was a John Buchan romantic, warm, loyal and generous to his comrades in great enterprises.

One of our most important jobs was to put together the winter crew list. At the end of the season the whalecatchers of the three expeditions, *Venturer*, *Harvester*, and Leith Harbour, as many as thirty-five in total, were laid up at Leith Harbour to be dry-docked, painted and repaired in preparation for the following season. The station itself had to be overhauled as well. That required a winter crew of 400. Some of the specialist tradesmen, like riveters and boilermakers, were brought down specially for the winter, but most of the crews were recruited from the whaling expeditions. The whaling contract had its notorious Clause K, which obliged every whaler to overwinter if required. That was not easy to enforce. A man who overwintered would sail south in the late summer of 1948, serve the 1948/49 season, go ashore at Leith Harbour for the winter, and return to his expedition for 1949/50. When he got home in May 1950 he would have been away for a year and a half, perhaps longer. Some of the older whalers had been separated from their families for five or six years during the war and were not keen on a further spell. Somehow or other by persuasion and mild threats we managed, in that year as in others, to get together the numbers and skills needed. Earnings were good for overwinterers, both in basic pay and overtime. The tax laws helped too. Both Norwegians and British were exempt from home tax if they spent a calendar year abroad, and long voyages were still an accepted pattern for merchant seamen. The British merchant navy signed its crews on two-year articles, and among our Norwegians a tour of two seasons plus a winter in the Antarctic was quite normal. We had a number of stalwarts, among them some particularly valuable catcher chief engineers, who made a practice of wintering in alternate years throughout their careers, positively looking forward to the peaceful routine of the overhaul months.

That season was one of *Venturer*'s best. The production of 169,000 barrels was well behind the 1947/48 postwar record of 203,000, but it was never reached in subsequent seasons. Blue whales were still quite common and strongly preferred. A blue whale was as good in size and oil yield as two fins. 643 blue were taken that season against 1,201 fins. It was the last big blue whale year. After that the catch of blues went down dramatically as overfishing depleted them and they became increasingly rare until finally given total protection by the International Whaling Commission. *Harvester* working in the east turned in 147,000 barrels, supplemented by 16,000 barrels of sperm oil. The expeditions often had two or three weeks of sperm fishing before the baleen whale season started. Sperm whales congregate and mate in warmer waters but in the summer months many of the bulls migrate to the Antarctic. These were hunted by the whalers. They were fairly easy to catch, but they could strike back. A harpooned and wounded whale often turned on the catcher, and its impact could damage the propeller or shaft, putting the boat out of action. So there was always the risk that a catcher would be disabled and have to go back to South Georgia or Cape Town at slow speed or under tow to get its propeller replaced or shaft straightened, with consequent loss of precious fishing days in the baleen season.

The oil of the sperm whale is not edible, unlike that of the baleen whales, and is used for lubrication and manufacturing. It fetched a price at that time a little lower than whale oil. Part of the sperm whale's oil is contained in a reservoir in its head (the 'case'). This pure and almost colourless oil was highly prized by the old American whalers, and it was carefully baled out from the head in buckets. In the modern industry all sperm oil was mixed together, as there was no longer any premium for case oil. Sperm oil and whale oil could contaminate each other and had to be kept carefully apart in the tanks.

An occasional bonus from sperm whales was ambergris, the waxy substance sometimes found in the stomach of sperm whales. Ambergris seems to accrue round beaks of squid which the whale has not been able to excrete, giving the stomach protection against abrasion by them. Robert Clarke, who later became a world authority on the sperm whale, was a government inspector on *Harvester* in the 1952/53 seasons. One day he spotted a suspicious lump in the guts of a sperm whale. He got the flensers to open it up, and it turned out to be a large lump of ambergris weighing 926 lb, one of the biggest weights ever recorded for a single piece. Robert gave himself the credit for having rescued this treasure. The chief officer maintained that the flensers were always looking out for ambergris and would certainly have found it two minutes later.

Ambergris has a mild sweetish smell. It has a particular value as a fixative for perfume and was sold by the pound for a very high price. Our piece was worth something like £25,000. We found a piece of similar size in a later season, weighing over 800 lb. I had it photographed as it came ashore at Liverpool and the picture came out in the *Echo*. That brought down on my head the wrath of Bill Hollyer, our partner responsible for sales. He pointed out forcibly that ambergris had a very limited market. Buyers would hold off when they saw that we had brought back enough to supply the world demand for three years. That chunk did go rather slowly, but I doubt if the price was affected by my photograph. We parked it in the cellar of our offices in Bernard Street, Leith, hacking off a bit from time to time when a buyer appeared. Its early promise was not fulfilled. As we dug further into it more squid beaks and other impurities appeared, so much of it turned out to have very little value.

Harold and I came back to South Georgia with *Venturer* at the end of the season and got one of the catchers on its way home to drop us off at Rio, from where we flew to London. With the arrival of *Venturer* and her catching fleet Leith Harbour was overwhelmed by activity. The catchers had to be moored and laid up in rows beside the jetties. Repair lists had to be completed and sent in to the winter workmaster. The catcher crews transferred to their cabins on *Venturer* for the voyage home, while the overwinterers came ashore to find their billets on the station. The secretaries lived in a turmoil of last-minute winter placings, completion of pay sheets, taking in and distributing mail, completion of voyage crew lists. The whaling season at South Georgia, which normally extended longer than the pelagic season, was still under way, but meat meal and bone meal from the station were already being ferried round by barge from the store sheds beyond the plan to be loaded on the transports. The same scene would be repeated two or three weeks later when *Harvester* arrived from the eastern grounds.

The winding up of the season at Leith Harbour gave people the chance to see old friends, catch up with gossip, and, thanks to the stocks of liquor allowed to gunners and masters, take a drink together. It was also a somewhat dangerous time. The moored catchers were connected to each other and the quays by planks and duck boards, often made slippery by snow. A careless step could put someone into the water, sometimes with fatal results.

In two or three days everything was sorted out and the floating factory manager saw with relief the station chimneys merging into the ring of mountains as the ship sailed out past Cape Saunders. During the voyage home there was plenty of time to clean the ship from the grime of the whaling season and scrape and paint all the deck houses. The factory machinery was likewise thoroughly dismantled and cleaned. But the whalers were also able to get some warmth in their bones from the passage through the tropics. On fine nights films were set up on the plan deck. Irving Berlin was played under the stars.

I was already back at work in Liverpool when the factories arrived to sign off the British crews and discharge in Gladstone Dock, which had the only berths big enough to take our ships. The oil was easy enough to take off. It was pumped into barges and delivered to Unilever's plant at Bromborough up stream. The meat meal was more difficult. There was no proper cargo hold on the factories, so the meal was stowed in some of the oil tanks below the factory deck. It was a slow painful business to hoist out the sacks one or two at a time on to the factory deck and then gather them in slings to be swung ashore by crane or ship's derrick. The dockers exploited these difficulties to their own advantage. So the ship had to spend a tedious fortnight in Liverpool before it could finally sail for Tønsberg.

The annual descent into Liverpool of several hundred whalers, awash with money, produced a considerable stir even in that boisterous port. Pubs doubled their stocks, girls dressed up, and taxis, the ultimate symbol of affluence, stood in ranks on the quay. Fortunately the final payoffs, with complicated bonus and overtime calculations, were not available when the whalers signed off, so they were only paid a limited advance. That ensured that, however attractive the joys of Liverpool, most of their earnings were still intact when they got home. They could generally be left to their own devices, but on one occasion I had to look after three Norwegian ex-overwinterers who had to be transferred from one of the transports in Liverpool to a floating factory then discharging down the Thames, for passage to Norway. They had been put on the train at Liverpool and reported as planned to our shipbroking office in St Mary Axe, where I was to pick them up. But our office was quite unable to control this drunken invasion and they disappeared into the City. Eventually after considerable searching I tracked them down, one trying to order a new suit from a smart tailor, the second persuading a bank to give him some money, and the third wandering down Leadenhall Street in a daze. A bizarre eruption from the Antarctic into the ordered world of finance.

The floating factories sailed into Husøflaket, their summer

moorings outside Tønsberg, in the first days of May, often in brilliant warm sunshine. The usual fleet of motor boats came out to accompany them on the last two miles up Oslofjord, while families thronged the jetty waving as the anchor was finally dropped. New babies were proudly exhibited, the February fruits of the previous year's homecoming. The whalers piled ashore, their whaling clothes cleaned and neatly stowed away in kitbags, but still assailing the sensitive noses of their spouses with the odour of old whale. *Tønsbergs Blad* was filled with interviews and writeups of the season's experience. It was the return of spring.

It was a matter of honour that all the pay sheets should be completed within a week of signing off. The Tønsberg office worked all hours calculating bonus earnings and overtime to get the final pay cheques of 250 whalers from each expedition handed over on time. That was not quite the end. Correct bonus payments could not be worked out until accurate outturns from the ships had been established. A small part of the bonus was held back and paid out as the 'rest part' on final figures in the autumn.

I spent a year in 1950/51 in Norway at the Tønsberg office under the stern eye of Salvesen's representative, Onésimus Andersen. Onésimus had been a stockbroker and an army reserve captain. He had visited South Georgia, got to know Leganger Hansen, and served a season as secretary there. That brought him, in 1927, to run our Tønsberg office. Onésimus had seen through the boom of the twenties and the subsequent shakeout, and his knowledge of the Norwegian whaling scene was invaluable. He looked after the buying of whaling equipment in Norway, the harpoons, grenades, guns and other tools of the trade supplied by local companies which had grown up around the industry. He recruited the Norwegian whalers, working with Max Harper Gow, and later myself, on the placings of the gunners and key officers. Onésimus coped sympathetically and firmly with all comers, whether it was the gunner who called in to drink coffee and lay on the line what he wanted for next season, or the abandoned wife, furious that her husband had stayed a further winter at South Georgia. Some of the whalers came in their best suits with briefcases, smarter than bank managers. Others called in more casual dress, and the briefcase might be bulging from a visit to the Vinmonopol, the government drink shop conveniently placed a little way down the street from the office. That was Nedre Langgate, with its row of solid white painted wooden houses, once the homes of previous generations of whaling owners, among them Svend Foyn, now the somewhat shabby offices of whaling companies.

Most of the whalers were local, from Tønsberg or Nøtterøy, the island that runs south from the town into Oslofjord. There was a strong contingent of catcher people, including some of the best gunners, from Onsøy, over on the east side of Oslofjord, who came across to visit via the Horten-Moss ferry. We also had a group from Aaseral, in the hills above Mandal. Lord Salvesen had built a large shooting lodge on the moors there to complement his Lorimer-designed house in Mandal. Harold had a small 'hut' close by. He introduced the crofters of Aaseral to whaling. They were not seamen, but their country skills made them excellent plan workers. As good peasants they were tenacious in looking after their interests. I would find an aggressive Aaseraling on the phone from the south, telling me forcibly that he would on no account go whaling again without two grades of promotion and a guarantee of no overwintering. But come October they mustered again for the new season.

During my time in Onésimus's office I took on the job of recruiting tradesmen for the Leith Harbour winter crews. Most of the crews came from the three expeditions, but they could not cover all the specialised trades needed for a large ship repair yard. We had to have riveters, welders, boilersmiths and a number of other tradesmen not required on whaling expeditions. They were taken on in February and sailed south on the transport ship of our neighbour station Husvik, to arrive in good time for the winter. They served the following season at Leith Harbour and returned the next May with the whalers. There was no difficulty in finding people ready to do that tour, both British and Norwegian. In choosing them I had no way of judging their skill and had to rely on record and references. I was caught on one occasion, when a very presentable riveter persuaded me to take as his holder-on a pal who he said had worked with him for years. When they arrived at Leith Harbour it was discovered that his friend had never seen a rivet. But he was an expert barber, and later became a useful steward.

At that time we got involved in lengthy negotiations with the Norwegian companies on pensions for gunners. This seemed a humdrum matter, but it was closely bound in with the delicate relationships between the gunners, the companies and the Norwegian government. There was a 'gentleman's agreement', dating back to the 1920s, between companies not to steal each other's gunners. That was generally observed, though Harold pointed out to me that gentleman's agreements are usually not agreements and not made by gentlemen. There had been some cooperation between the companies in the 1930s on rates of pay and bonus, and this was formalised after the war in yearly negotiations with the Gunners' Union. The Norwegian government wanted a hold on the gunners in two ways. It was out to stop any new nations coming into the Antarctic, and it saw as its most effective weapon a ban on Norwegian gunners taking employment with those countries. It also disapproved of high earnings. It was running a tight incomes policy, in conjunction with the unions, which, to be sold to the public, had to be given some egalitarian ideology. The gunners did not like this. Their relative earnings were pulled down and opportunities for bettering themselves elsewhere were blocked. The term 'bond slaves' was sometimes brought out. While a few defied the law and moved abroad, to the Dutch or Onassis, most of them accepted the position. But the companies were nervous about too much outside pressure on gunners, who were after all their key people. The pension scheme was an effort to give them more recognition without upsetting Norwegian government policy.

The negotiations on this were complicated. The pension plan was intended to eliminate any other payments by the companies outside the negotiated season's contract. Companies were reluctant to agree when they were suspicious that by signing they might be put at a disadvantage in relation to less honest competitors. Harold adhered scrupulously to his agreements, but was sceptical about other companies. The pension levels agreed looked substantial at the time, but when they came to be paid out twenty years later were so reduced by inflation that they were little more than pocket money. I found out many years after the end of whaling that one of the companies which had signed the agreement had still run its own private pension scheme for gunners.

Despite the erosion of their position gunners continued to be the aristocracy of Tønsberg and Sandefjord. On the whole they were careful spenders, but they built themselves fine houses with jetties and grand motor boats. One of them had a nice grey painted fence round his garden, revealed to me innocently by his young daughter as having been done with *Venturer*'s hull paint. Hans Abrahamsen on the occasion of *Venturer*'s 200,000 barrel season in 1947/48 was given an immense motor boat by the company, the most potent symbol of prestige he could have.

When the first labours of restoring the whaling fleets after the losses of war had been completed the main concern of Antarctic whalers reverted to the whale stocks. In the last seasons before the war catching had reached about 40,000 whales a year. Although five years of respite during the war had allowed some recovery no one with any experience seriously believed that this high level of catch could continue. When the new International Whaling Convention was signed in 1946 an agreed total limit of 16,000 'blue whale units' was written in. The blue whale unit was an attempt to equate the different types of whale in a standard measurement. One blue whale was reckoned to be the equivalent of two fins, two humpbacks, or six sei. This arrangement was intended to encourage whalers not to concentrate on the larger blue whales, which were clearly being overtaxed, but to spread their catching to the more abundant fin whales. It certainly helped to take away some of the strain on the blue whales. But, as it turned out, it was already too late to help the blues, whose catch in 1948/49, at 7,625, was little more than half the catch in 1938/39. After the war and until the final collapse of whaling the preponderance of catch was of fin whales.

The catch limit of 16,000 blue whale units compared with catches of 30,000, 25,000 and 21,000 units for the three last prewar seasons. The three scientists who recommended it, Remington Kellogg of the USA, Dr N. A. Mackintosh of the National Oceanographic Institute and Professor Birger Birgersen of Norway hoped that this was a level that could be sustained for the future, though each of them, looking back at the records of the past, had his doubts.

The setting, for the first time, of an overall quota helped to limit a new expansion of the industry, as well as giving some protection to the stock. With whale oil at £100 per ton a whaling expedition to the Antarctic looked an attractive venture and there were several nations eager to come in. Apart from Holland and the USSR, who succeeded, there were plans in Germany, Italy, Sweden, Argentina and the USA. It was clearly going to be much more difficult to establish a profitable operation in competition for a share of a fixed quota than if the field had been clear.

For those in the industry the fixed quota made it vitally important that no more expeditions came into the Antarctic. The eighteen fleets operating in 1948/49 were quite enough to take 16,000 units during the operating season. Each extra expedition would reduce the catch of all the others by 5% or so. When the price of whale oil came down from its scarcity level that difference could convert profit into loss. So any aspiring new entrants were countered by diplomatic pressure from the Norwegian and British governments and by attempts by the companies to buy them out. Argentina, in the shape of Alfredo Ryan, the ambitious owner of our neighbour station at Grytviken, ordered a floating factory from Harland and Wolff in Belfast, the *Juan Peron*, which was to carry the glory of his country and its dictator into the Antarctic. Fortunately the financing failed and the ship was left to sail as the biggest tanker in the Argentine fleet. We took over two of the whalecatchers he had ordered from Smith's Dock, virtually sister ships of the boats we had been building there. Harold's judgement of Ryan was expressed in the name we gave to one of them – *Southern Gambler*.

The other serious competitor proved unstoppable. In 1950 the shipowner Aristotle Onassis put together a whaling company to convert a war-built T2 tanker into a floating factory, *Olympic Challenger*. The Norwegian whaling companies, led by Anders Jahre, did everything they could to prevent Onassis's plans from being carried out. Onassis had an American associate who had been involved in whaling with Jahre before the war, and Jahre appealed through him to Onassis to give up the venture. But Onassis was intent on what he saw as a lucrative speculation. He was not concerned about the effects on other Antarctic operators, and he turned down an offer by the Norwegians to buy him out at a handsome price. The *Olympic Challenger* conversion went ahead, with Onassis securing the support of one of the prewar German whaling companies. The fleet was converted in Germany and supplied with German crews, except for a few Norwegian gunners who joined it in defiance of their national law. As manager Onassis succeeded in getting the notorious Lars Andersen, who had collaborated during the Nazi occupation of Norway. Onassis had registered the ship under the Panama flag, and Panama was a member of the International Whaling Commission, so his expedition was required to operate within the fixed overall quota. In practice he ignored it. Although he tried to keep this from the public eye, reports very soon started to come through of the expedition fishing outside the season and taking large numbers of undersized whales. Eventually, as will be seen later, his crimes caught up with him. Meanwhile they cast a blight on the international cooperative efforts needed to conserve the whale stocks.

The second yearly meeting of the International Whaling Commission, the body set up by the International Whaling Convention, took place in Oslo in summer 1950. I was still working in Tønsberg and was able to attend it with Harold. He and Rupert Trouton of Hector Whaling were the appointed industrial advisers to the British delegation. The signatories to the Convention included all the Antarctic whaling countries and a number of countries whose participation in whaling was quite small, such as Brazil, New Zealand and USA. The USA, as the leading founder of the Convention, was also the strongest champion of conservation. Others, not engaged in the Antarctic but with their own domestic whaling industries, found it more difficult to take the pure conservationist line. Canada argued consistently for the reduction of Antarctic catches, but had at the same time to plead for a dispensation for her stations to catch smaller whales than the international minimum size, since she claimed that the survival of whalers in the northern hemisphere depended on it. Such situations, where economic requirements clashed with ideals of conservation, formed the pattern of international whaling negotiations. Harold sensed more than a little hypocrisy in them, and did not hesitate to say so publicly.

The overall quota remained at 16,000 till 1953, being reviewed annually by the Scientific Committee, which on the basis of the scanty evidence then available tried to advise the Commission of the state of the stocks. While the preponderance of world catch was in the Antarctic, there were many land stations all over the globe, most of them dating from the early Norwegian diaspora, others set up after the Second War when oil prices were high. In the southern hemisphere these fished the same stocks as the Antarctic whalers. So there was perpetual tension between the shore-based whalers and the floating fleets. The shore stations would not accept any catch quotas on the grounds that their catch was self-limiting. A station with an operating radius of no more than 200 miles, however hard it fished, would not make much impact on the pelagic stocks of the whole Antarctic. A more convincing argument was that shore stations were much more vulnerable than floating factories. If catches failed they could not move to a different area. The land stations, except for Peru and Chile, accepted the six-month season that had long been applied at South Georgia. They naturally fought for any reduction in whaling effort by the Antarctic pelagic fleets which would bring them more whales.

The humpbacks were a particularly sensitive issue. They were easy to catch and stocks could quickly disappear unless there was control. In the Antarctic they had been protected since 1938. When Australian stations started to catch them after the war and it seemed that stocks had recovered, the Antarctic whalers claimed their right to fish them as well. So a separate quota of 1,250 humpbacks was set in 1949, within the overall quota. That, with subsequent variations, still left enough whales for the Australian stations for many years, though eventually overfishing by everyone, including a great illegal Russian haul in 1960/62, knocked them out. The Australians and South Africans, whose stations were supplied from the Antarctic whale stocks, claimed that the whales which came up and down their coasts were their property and should be reserved for them. The Antarctic whalers, against that, argued that the whales taken in warmer waters often yielded only half as much oil as the same whales taken in the course of their feeding season in the Antarctic, so it was grossly wasteful to catch them when they were so thin. The vulnerability of the humpbacks was shown by a French-Norwegian venture off the Congo in the early 1950s where 3,000 humps were taken in two seasons, and the local stock trekking down that coast was cleaned out.

A comedy part of the Commission's meetings was the annual debate about indigenous catches, still I believe a feature of IWC proceedings. The right whale of the Arctic was now very rare and fully protected. But the Canadian Eskimos had always taken a few of those for oil and meat, so their privilege had to be preserved, in defiance of the high conservation principles of the Canadian government. The South Africans claimed that they too were feeding indigenous people with whale meat and should have similar concessions, but they got no support.

Among the delegates our own Fisheries Secretary, who headed the British contingent, used to stand out in competence, showing the high standards of our civil service. He often took the lead in steering proposals through and drafting acceptable amendments. But we sometimes felt that, true to the high-minded British tradition, he was trying to act as a little friend of all the world rather than, like everyone else, pressing the claim of the whaling industry of the country he represented. The Norwegians were well-informed but not versed in international negotiation. The Commonwealth countries, Canada and Australia, were remarkably bureaucratic in their outlook, floundering in matters of procedure to the prejudice of important issues. The Japanese, with an expanding confident industry behind them, left the initiative to others and accepted new regulations as they were proposed. Whether they always carried them out is another matter. The Russians had the advantage that the captain of *Slava*, Solyanik, was also a delegate and could speak out freely. Solyanik was from Vladivostok, swarthy and black-haired, decorated as a Hero of the Soviet Union. While the Western delegations were generally cool towards the Russians, Harold determinedly made friends with Solyanik, 'a real man', though he had no illusions about his whaling ethics. He and Harold were both members of the Technical Committee, one of whose jobs was to scrutinise reported infractions of the Convention such as catching undersized whales or fishing outside the prescribed season. Harold reported that sometimes, when Solyanik was about to deliver one of his whopping lies about illegal catch in which *Slava* had been observed red-handed, he would cast a conspiratorial wink at Harold on the other side of the table. Harold had some words of Russian, picked up when he briefly commanded a company of White Russians in north Persia during the First War. Occasionally he could catch the drift of what the Russian delegates were saying before it was translated and that gave him their respect.

Solyanik went on to command the first of the new floating factories which the USSR built from 1959 onwards. But his distinguished career later crashed. He was accused, improbably, of having used money intended for crew welfare to build himself a private swimming pool on the ship and he was exiled to command a tug in the Black Sea. Later he emerged again as master of a shrimp factory ship involved in some conflict with a local boat in the Gulf of Carpentaria, north of Australia.

Harold also diverted himself criticising the scientific evidence that was being brought forward on the state of the whale stocks. The biology of whales had been well studied, but there was no reliable way of telling the age of the whales caught, one of the elements necessary for stock assessment. An early method was to count the number of *corpora lutea* of pregnancy in the females. That had deficiencies, as did the counting of yearly ridges of growth on the baleen plates. Eventually the best indicator was found to be the wax ear plugs, which added a new layer each year. As regards general abundance, that was only being estimated by extrapolating sightings from the bridge of *Discovery* on its cruises through the Antarctic. Harold was very scathing about the figures produced, and the brunt of his sarcasm fell on Dr Mackintosh, the principal scientific adviser to the Commission. This gave him the reputation of being totally opposed to any conservation measures. I had to try to persuade worried delegates that Harold realised as much as anyone that the stocks were going down and wanted them conserved, but that he couldn't accept arguments that he thought were intellectually disreputable.

I moved back from Norway to Leith at the end of the summer of 1951. Soon afterwards Harold and I set out for another season in the Antarctic. That was for me the most valuable possible learning period. I was by now fluent in Norwegian, knew well most of the key people, and was sufficiently at home on the expeditions to be able to concentrate on the many operational questions which Harold was wrestling with. We spent much time on the factory deck studying processes and working out ways to improve them. I filled notebooks with details of the production plant and the passage through it of the whale material to be broken down, cooked and recovered as oil and meat meal. The initial fairly simple layout of the factory decks on *Venturer* and *Harvester* scarcely lasted their first operating season. New centrifugal separators were brought in to increase oil yield, and there were continual fresh ideas to improve capacity for both oil and meat meal. Existing plant had to be moved and crammed even tighter to allow space for the new machines. Newcomers trying to find their way from one end of the factory to the other lost themselves in this maze. It needed considerable study to follow the sequences of pipes and tanks which linked the processes. One of our chemists, having spent many hours trying to decode the systems, announced that he had followed a pipe twisting through the factory and found that it simply went round and joined itself again. The complexity was worth while. The ships had excellent results in oil output per blue whale unit compared with their competitors. But we often thought wistfully of larger or less well-equipped factory ships, some so spacious that a small car could be driven down the centre aisle.

We visited both the floating factories that year and it was most interesting to compare their methods and try to transfer to each the benefit of the other's experience. I had plenty of time to go out on catchers, spending hours on the bridge with the gunner looking out for whales, which of course I was the last to spot. The gunner was continually on the bridge during the hours of daylight. In mid-summer there was little darkness and he had to snatch what sleep he could when the boat was steaming to pick up the day's catch of whales or tow them in to the factory. Many gunners slept on their cabin settees in their clothes during the season, and left their bunks untouched. Occasionally there were days of fog or bad weather when arrears of sleep could be made up and there was time for whale gossip in the mess over cups of strong Norwegian coffee. The Norwegian steward of the catcher was always a seasoned cook, providing an excellent table, although catchers had a lively movement in heavy seas and cooking could become hazardous. The traditional dish, when the pots were sliding off the galley stove and the coffee cups cascading down from their hooks on the mess deck head, was *stormsuppe*, a mixture of everything eatable in the form of a thick stew, served in a soup plate and consumed spoonful by precarious spoonful.

The catch of the two expeditions was recorded in the 1951/52 season as being fifth and sixth in order among the fifteen European expeditions. (The South African expedition was included because it had largely Norwegian crews.) That was somewhat lower than the ranking we expected to get. The Norwegian and British companies were intensely competitive in comparing catch figures with their rivals. We managed to hold a high ranking, particularly in the early postwar years, though we were beaten by *Balaena* which had some spectacularly good seasons, led by a first-rate manager and team of gunners. The Norwegians attributed their poorer results to government restriction on new investment in whalecatchers after the war. As time went on the Norwegians caught up with the British and there seemed to be no marked difference between the two groups. There remained considerable variations in catch, depending on the whale and weather conditions in the area chosen for fishing. We, from our base at South Georgia, tended to choose the nearby Weddell Sea for one of the expeditions, whereas *Balaena* and the Norwegians, going south via Cape Town, were attracted by the 'middle field' south of there. That probably over the years produced rather more stable catch and somewhat better weather than our home waters. Our own operations were in safe hands. Harald Myhre served as manager of *Venturer* from 1949 till the ship's sale in 1962, and the two leading gunners, Hans Abrahamsen and Arne Mikalsen, were outstanding among their fraternity. We did not win all the prizes, but we got a good share of them.

On my return from the Antarctic in 1952 I settled in Leith to take over the Crew Department and was subsequently handed stores and production. My cousin, Max Harper Gow, continued to have general responsibility for the whaling operations under Harold. There was plenty for me to do as well in the crewing and storing of our deep sea and short sea tramp ships, and Max and I between us took over the operational responsibility for them after Iver Salvesen's death in 1957.

Our Crew Department in Leith, as in Tønsberg, had a seasonal pattern of activity corresponding to the requirements of the whaling fleets. During the spring and summer there was intense work paying off the whalers when they came home, forming the crew lists for the following season and finally despatching the expeditions to the Antarctic. The Leith Harbour contingent sailed in late August, to be down in time for the start of the South Georgia season on 1 October. The floating factories left in mid-October, with the pelagic season starting in December. Having had much of their summer overhaul done by their Norwegian crews in Tønsberg they would sail over to Newcastle in August for dry-docking and any big repairs or alterations. In October they loaded their stores in Newcastle, took on the British crews, and crossed to Tønsberg to embark the Norwegians. From there they set off for South Georgia, calling at Aruba on the way to fill up with fuel oil.

The weeks before the factories sailed from Newcastle were particularly hectic for our Crew Department. The whalers were summoned from all parts of the country, sixty or seventy a day, to get their documents prepared and for a medical examination, usually from the doctor who was sailing with the expedition. The operation was lubricated by convenient pubs in Bernard Street and the Shore.

The mix of Norwegians and Britishers, as we called them, had by now settled down. Among the officers Norwegians dominated. Most of them had made whaling their career and shaped their professional training accordingly, whereas British officers came in more as an interlude in their merchant navy life. The engine rooms on the factory ships were wholly British-manned, starting with the chief and second engineer, who were required by law to hold British certificates. British engineers also found places on the catchers, many of them from the marine communities of Tyneside. For deck and plan work in the factory staff there was a full mixture. Particular strength came from the Shetlands and the outer isles of Scotland. These were men who had served the sea for generations and took in their stride the hardships of whaling. They provided a backbone no less sturdy than their Norwegian counterparts.

We would have liked to get broader parity by training up some British gunners, but there were great difficulties in doing so. There were few suitable candidates, and fewer who would have been ready to serve the necessary apprenticeship as a catcher second mate and mate and achieve fluency in Norwegian. We had a single British gunner when our whaling finished, with no other aspirants in sight.

One of my jobs was to recruit doctors. We carried a doctor on each of the floating factories and one, sometimes two, at Leith Harbour. We also had to provide a doctor for the winter there, taking turns with the Husvik station. Just after the war it was very difficult to find doctors to go whaling. Those we got were either newly qualified men, enthusiastic enough but without the experience which was needed to take sole responsibility for 600 men in virtual isolation from the rest of the world, or persons with a history. Sometimes that history did not emerge until they were away on the whaling grounds beyond possibility of recall. Their failings were a rich source of whaling anecdote – drink mainly, but also drugs and mental collapse. Fortunately by the time I came into the picture the worst was over. We somehow managed to recruit for our needs without any disastrous failure, and with eccentricities well within the tolerant ambit of whaling. Two of them served for many seasons and became pillars of our society, Dafydd Richards, a wise and humorous Welshman, and Ian Mackintosh, fond of whisky but a true healer.

One weakness to which a number of whaling doctors were prone was writing books about their experiences in the Antarctic. An early one was by a doctor on a Norwegian floating factory in 1916. He recounted having found among his stores on the ship five complete sets of midwifery instruments. No women had ever sailed on the ship, but the instruments had been put on the stores requisition list when the ship was first commissioned, no doubt in obedience to some government regulation, and the order had been automatically repeated every subsequent season. The same doctor called on a

British factory ship in the South Shetlands, almost certainly one of Salvesen's, and found that the doctor was permanently drunk.

The most notorious publication was Dr R. B. Robertson's *Of Whales and Men*, referred to earlier. Robertson captured the atmosphere of whaling very well, but much of what he wrote was fiction. His book was first published in the USA, where it became a best seller and was serialised in the *New Yorker*. Harold, brandishing threats of libel suits, succeeded in getting the most offensive parts removed before it was published in the UK. It was not only Harold and his partners who resented being pilloried as ruthless bible-punching capitalists exploiting an oppressed proletariat. The whalers themselves were equally indignant at the treatment they were given. Our chief engineer of *Harvester*, Charlie Crockett, veteran of prewar factories and Atlantic convoys, was most offended to find that he had told Robertson that for twenty-five years' service he had been given a duplicated letter of thanks from the owners and a 17/6d alarm clock.

Despite the improbability of most of what he wrote, Robertson's book will no doubt remain a source of whaling history. I found him quoted in a reputable biography of Shackleton as having met on a Salvesen ship a whaler who had been at Stromness on the morning thirty-four years earlier when Shackleton and his comrades arrived there after crossing the mountain barrier. I had to write to the author to say that there there was no such whaler with us in 1950 and that Robertson could not be considered a reliable authority.

Though we learned to be wary of any doctor applicants who seemed to have literary interests, I did take on one who had already made a reputation with humorous books about doctoring and clearly wanted to add whaling to his list. But on the way over from Newcastle to Tønsberg he was seized by a mysterious stomach pain which he refused to have examined. We were forced to discharge him in Norway and find a substitute at short notice. He had evidently had second thoughts about the responsibilities he was expected to carry as the price for collecting literary material.

The doctor always livened up the expeditions. He was the one person on board who as an outsider could stand away from the whaling obsession and bring in some fresh air to relieve the tensions that could build up over the season. One of them, a railway buff, got a fierce argument going in the top mess on the number and names of the stations on the line between Melrose and Hawick. It raged for weeks and was not resolved until the ship got back to Liverpool and someone had procured a railway map.

The doctor's professional responsibility was considerable. His crews had all been passed as fully fit when they signed on, but some always slipped through with undetected illnesses. The whalers were remarkably free of accidents at work, considering the risks. On the catchers there were all the hazards of rough sea compounded with the immense strain put on ropes and equipment in the handling of the whales. On the factory decks slippery piles of material were crossed by hooks and wires under tension. Saws and flensing knives, dealing with the moving masses, were another source of risk. The whalers followed habitual routines of safety and major casualties were rare, though there were the inevitable cuts, bruises and strains, and sometimes broken bones to set. I was on board *Venturer* in 1952 when one of the few fatal accidents occurred. It was on a catcher when a whale was being pulled in by the whale line under tension. The harpoon suddenly flew out and hit the gunner in the shoulder as he stood on the gun platform. He was brought back to the factory and the doctor operated, but he could not be saved.

If the doctor could not treat a patient properly on the floating factory his instinct was to evacuate him to where he could get better treatment. But the whaling manager would resist this. A whalecatcher detached from the expedition to take a patient to Cape Town or South Georgia would forfeit a lot of fishing in its absence. Besides, as the doctor too would realise, a ten-day trip in a catcher through rough seas would do the patient no good. So hospital cases usually had to stay on board, perhaps being transferred by a convenient tanker to Leith Harbour, where there was more room for nursing. The worst people to deal with were mental cases, who sometimes needed to be closely watched. At the end of one season Max Harper Gow sailed with *Polar Maid*, one of our small tankers, from Leith Harbour to Montevideo. On board was a mental case for repatriation, with two attendants. When walking on deck with them the patient slipped out of their control and jumped overboard. That proved infectious. Two days later the second engineer emerged from the engine room on to the poop deck, where the chief engineer was sunning himself, clasping a heavy spanner. He gave the chief a courteous goodbye and went over the rail into the sea. Both jumpers were lost. There was not much chance of survival in those cold waters, particularly if the man concerned was bent on suicide.

The employment of our whaling crews centred round the whaling contract. Joint negotiations between the British and Norwegian companies and the British and Norwegian unions had been initiated by Harold before the war and were strengthened after it. Every summer there was a full-scale negotiation at the offices of the Norwegian Whaling Employers' Association in Sandefjord to decide on the new season's contract. The negotiation became a set piece, with an initial presentation by the unions' leader to make their claim for an overall increase, eloquently and passionately argued for the benefit of their audience of union representatives. The owners' reply followed on the same level and the meeting then got down to studying detailed proposals from the unions for amendment of the contract. The show was run mainly by the Norwegian Association, which provided massive statistics to support its negotiators. A Norwegian owner led for the companies, and the head of the Norwegian Seamen's Union spoke for the unions. But Harold played a leading part. A number of the clauses had been introduced or drafted by him and, as noted earlier, he had introduced the three-year average bonus basis. Harold made sure that the British seafarers' unions were an integral part of the negotiations. Their presence emphasised the permanence of British crews on British expeditions. With whaling pay and bonus rates continuing to be far above the British merchant navy, they were unlikely to press as hard as the Norwegians and could be a restraining influence. The British unions had their own agreements for merchant navy seamen under the National Maritime Board, for many years an effective and respected body. Harold got a National Whaling Board set up as the equivalent in whaling, though it did little more than endorse formally the agreements of the Joint Negotiations. Harold also had the idea of forming the Whalers' Consultative Committee, a small body of owner and union representatives which worked to get agreement on some of the detailed claims, apart from pay, before the main negotiations. There were twelve grades of pay groups below officer status, ranging from chief steward in I, through skilled tradesmen in IV and V past unskilled whalers in VIII and IX, down to first voyage boys in XII. As in all grading schemes there was continual pressure upwards. Why should a boiler emptier in VI be lower than a 2nd separator man in V? The WCC was quite a good body to argue out such issues in an objective way, though the Norwegian unions did not like it much because it suppressed some of the thunder they prepared for the main and public negotiations.

This negotiating system worked smoothly. Throughout the postwar period there were never any serious labour troubles. The yearly contract left companies and whalers free to make their own new arrangements each season, so there was no conflict over redundancies, while the custom of whaling, fortified by Harold's clause, cut out any demarcation questions. But the main guarantee of its success was Norwegian government policy. The Norwegian government worked hand in hand with the Landsorganisasjon (LO), the equivalent of our TUC. Pay settlements were kept under strict control. If there was no agreement between the parties the dispute went to arbitration and could end in a compulsory settlement by a government-appointed court. On one occasion the whaling unions decided that they wanted a 25% increase, and stuck to the claim, breaking off negotiations. It was agreed to refer the claim to a Joint Arbitration Tribunal with an independent chairman nominated by agreement. The tribunal awarded 12%, and the dispute was settled. Looking at the destructive anarchy of wage settlements in Britain we reflected how lucky we were.

There was one occasion of real conflict. In 1946 the Norwegian unions revived their old claim of 100% Norwegian manning, but limited to whalecatchers. This time there was no support from the Norwegian government or anyone else. Britain had after all poured out its life blood to rescue Europe from the Nazis, and most Norwegians were very conscious of their debt. The British unions were firm behind us, and the claim was dropped.

There were rarely disputes on the whaling grounds. The agreement provided for any conflicts to be resolved by negotiation at home after the end of the season. That nearly broke down on one occasion, when I found myself in the front line. I was at Leith Harbour at the end of the 1951/52 season helping with the end-season transfers and dispatches. My seniors at home decided to reinterpret the agreement for the floating factory men who joined Leith Harbour for the winter before the end of the land station season. These had previously got the 75% winter addition to basic pay from the day they landed. The new arrangement was to be that they went on to bonus for the Leith Harbour catch until the land station season finished on 31 March, and only moved to the winter addition from 1 April. They stood to lose on that, as catch, and so bonus earnings, was low at the tail end of the Leith Harbour season. There was a great protest, and the floating factory whalers refused to go ashore unless the new arrangement was cancelled. I had to stand firm and try to persuade them to work normally, referring the dispute for settlement at home. It would have been a blow to our carefully developed agreements and procedures if we had been forced to give way to a strike threat. It was a nerve-wracking position for me, still limited in my experience, and only able to seek advice from home through laboriously coded telegrams. Fortunately the union representatives on the spot were steady, and together we managed to persuade the whalers concerned to go ashore for the winter without my having to indicate any concession to their claim from the company. When the dispute came up at home the company gave way, as I always thought it should, and the old arrangement was reaffirmed. I felt we had been mistaken, not in promoting a new arrangement which in some ways was more logical than the old one, but in springing it on the whalers without warning or negotiation.

The organisation of stores and provisions for whaling operations was complicated and demanding. Everything required for the two floating factories for the season, and for Leith Harbour in both its whaling and winter repair yard capacity, had to be arranged in time to be loaded on the ships at the end of the summer. There was a scramble to get the lists from the heads of departments as soon as they arrived home and to send out final orders to suppliers as early as possible. One of the biggest items was rope for the foregoers and whale lines on the catchers. The foregoer was the lighter rope attached to the harpoon, and it was backed by a heavy whale line. Up to 1939 both foregoers and lines had been of manilla. The development of synthetic fibres produced the nylon foregoer, stronger and more elastic, and that was universally adopted. Our suppliers, British Ropes, had a factory in Leith. There, on the rope-spinning machines which had superseded the old rope walk, they produced all our foregoers half a mile away from Bernard Street. While whale lines would normally last a season, foregoers were subject to immense strain and wear and had to be frequently replaced.

Leith Harbour, in addition to normal whaling supplies, needed all the furniture of a shipyard: steel plates, non-ferrous metal supplies, machine tools and their spares, spare propellers and whalecatcher components. There was no just-in-time service for the Antarctic, so everyone tended to over-order to be on the safe side. With whale oil at £100 a ton, and Harold's general feeling that the whalers must have the best, we were perhaps too tolerant in controlling the demands of the expeditions. It was anyway difficult to do so at Leith Harbour, where for the first ten years after the war there was a large reconstruction programme of machinery and buildings, including a new floating dock. That was built on the Tees, then dismantled and shipped south to be put together again at Stromness. On later visits to South Georgia I found the stores packed with a superfluity of treasures which had been ordered in good faith and never used. We discovered on one of the tankers a supply of cooking pots which our catering superintendent calculated would last the ship, at current rates of consumption, for 108 years.

Stores supervision covered food as well, and I became increasingly interested in our catering arrangements. Frozen food was starting to find a place in ships' catering, though its scope was limited by the cold storage space available. We achieved a great improvement by buying our frozen meat ready butchered in flat packs, so leaving space in the stores for frozen peas and fruit. Salt fish was still an important item, carefully packed in wooden boxes with tinned liners. This came in the messes as '*baccalao*', fish stew with onions and potatoes. Wednesday was the day for it on the factories. By far the biggest item was potatoes. It was said that whalers would tolerate shortages of meat, butter, coffee, indeed of anything but potatoes, but if potatoes started to run out there would be mutiny. We bought our potatoes carefully, choosing the longest-lasting ones and shipping them in crates that would give plenty of ventilation during the voyage south through the tropics.

The last consignments were the most important, as they had to carry Leith Harbour through the winter. The best potatoes for the purpose were found in Ringerike, inland on the eastern border of Norway, so we took our late supplies from there. The chief steward had to keep a wary eye on potato stocks. We tried out an electrical chip fryer on one of the factories, to the joy of the British, but it did not last for long. The power demand on the ship's generating plant, and the enormous rise in potato consumption as chips caught on with the Norwegians as well, forced the steward to shut it down.

Fresh whale meat came on the menu, but as a special treat rather than regular fare. The best way of treating it was to cut a fifty-pound fillet from the back and hang it on the rigging for a few days to allow the oil, which gave it a slightly fishy taste, to drain out. It was then as good as anything that could be provided from the freezers.

Apart from the normal manning, supplying and operation of the fleets our main concern at head office in Leith was to put their catching and production efficiency ahead of everyone else's. By 1951 the postwar catcher building programme had been completed and we had a full complement of 14-15 knot boats. But we still had some of the single-boiler prewar catchers which could not match the rest. The Salvesen board led by Harold set its face against any new large investments in an industry which was starting to look precarious. The way out was in buying good second-hand catchers from other companies and in converting war-built corvettes. With their large size and power they far outclassed the Smith's Dock catcher, and their cost was a fraction of that of a new building. The best gunners got the corvette catchers, so there arose quite a gap between them and their less skilled colleagues on smaller boats.

Our main development to improve catching was asdic. The location of under-sea objects by sound waves had been worked out during the war and successfully applied against U-boats in the Atlantic. Harold saw its potential for tracking a whale during the chase and guiding the gunner to place the ship ready for a shot as it came up. Special equipment was designed by Kelvin Hughes with a dome containing the scanner that was let down below the hull and retracted when not in use. Over a few seasons the two floating factory fleets were converted to asdic. Each boat carried an operator, initially men trained on naval ships. We soon trained our own operators, who combined this with deck work on the catcher. The gunners were convinced that asdic reduced their chasing time before the kill and so increased their catch. Part of the benefit of asdic was simply that it frightened the whale and made it run. Whereas on the older, smaller, catcher the gunner exercised his traditional skill in watching the whale dive and trying to guess where it would surface, the large boats set out to keep the whale running at full speed near the surface until it slowed with exhaustion. But the skilled asdic operator was certainly much more than a whale frightener, and the best gunners worked closely with them. Asdic, however, had some drawbacks. The domes took half a minute to lower, and a minute and a half to take up again. After the whale was hit the dome could easily be damaged by the rope or by impact with the whale itself, requiring a trip to South Georgia and the loss of valuable catching days. A quick lift system was later developed, but it was too expensive for general fitting.

We were almost alone in adopting asdic in the Antarctic. That might suggest that we were optimistic in our faith in it. But it did seem to work for us. There were so many elements in catching success that it was difficult to isolate and accurately assess any one of them. Other fleets concentrated on different elements – speed of catchers, quality of gunners, radio communications, floating factory mobility. Any of these may for them have compensated for the advantages we thought we gained.

Aerial spotting was a similar aid to catching. As far back as 1930 the new *Kosmos* had carried a spotting plane. It was lost on its first flight. *Balaena* started off with a seaplane, but discarded it after the first season. Salvesen started thinking about the possibility in the early 1950s, encouraged by the development of the helicopter, which would allow take off and landing on a stern platform. The principal promoter in the company was Clifford Marshall, son-in-law of Noel Salvesen and a wartime flyer. He insisted that two per factory were necessary to give full spotting service and advisable for safety. Harold thought that one per ship was enough. Clifford put in his resignation which Harold, in character, accepted. The helicopter scheme went forward, with two aircraft for each factory, Harold eventually moving over to accept the arguments he had opposed.

In our enthusiasm for what they could do we did not appreciate the essentially delicate nature of helicopters. All went well in the first season, 1954/55. But on the way south the following season one of the aircraft flew into a tropical storm, lost its horizon and crashed into the sea, with the death of one of the pilots. In subsequent seasons there were further casualties to aircraft, fortunately without loss of life, which eventually put the whole operation in doubt.

The helicopters became an important part of the floating factory operations. The pilots and engineers were an alert team who worked hard to keep the craft in the air whenever the weather allowed. They cooperated well with the gunners, and at the end of each season produced impressive figures of the numbers of whales spotted by them and subsequently shot. Apart from the worry about their safety, which lay heavily on the expedition manager and the company managers at home, they seemed to justify themselves.

The helicopter crews were very different people from the whalers. They were used to the strict disciplines of flying and aircraft maintenance, but not to the laborious and austere conditions of the merchant navy, dominated by the sea. To bring them to the Antarctic at all we had to allow them the privilege of their own liquor stocks, putting them on the level of the manager and the doctor. That was envied by the rest of the crews. But the airmen were accepted as being sufficiently distinct to justify special treatment, and they never abused their position. Their cabins were at the stern of the ship, close to the helicopter pad and garage, and they formed a cheerful colony there, worshipping their own gods of the air.

We were presented in 1956 with a magnificent painting of *Venturer* in full whaling operation (see illustration). The original stern view showed the helicopter hangar shut and no aircraft outside. Our helicopter staff naturally felt insulted. The painting was sent back to the artist, who obliged by opening the door of the hangar and putting in one of the helicopters, while the second was shown setting off on a spotting trip.

The introduction of helicopters into whaling put into some people's minds the idea of shooting whales from the air. One of those interested was Colonel Blacker, the ex-Indian Army colonel who had invented the bomb-throwing Blacker Bombard for Home Guard use during the war. (I had lessons on it on the downs above Marlborough in 1942.) But shooting from aircraft could only be effective if the whale was killed stone dead and then floated so that it could be picked up by a boat and towed to the factory. No aircraft could attach itself to a harpoon line and reel in a whale. Blacker's weapon was no more practical than any other of the alternatives proposed from time to time to replace the explosive harpoon.

After the initial rebuilding and expansion of the fleets the number of Antarctic expeditions stabilised at around nineteen. In 1954 that was divided into three British, nine Norwegian, one South African, one Dutch, *Olympic Challenger*, three Japanese, and the Russian *Slava*. The total quota having been held at 16,000 up to 1952/53 was reduced in 1953/54 to 15,500. The increasing competition as the fleets improved their efficiency was shown by the shorter time needed to take the quota. In 1951/52 the 16,000 units then available were caught in 64 days out of a total permitted season of 97. The industry had quickly reached gross over-capacity, operating for little more than two months of the year. The Whaling Convention provided no means of combating this.

The only way capacity could be reduced and costs brought down was by cooperative efforts within the industry, and these were initiated by the Norwegians and British. The result of lengthy negotiations, started in 1952, was whalecatcher limitation agreements for 1953/54 and 1954/55. They covered all the Antarctic expeditions except the Russian *Slava*, which could not be persuaded to take part, and, of course, *Olympic Challenger*. There was natural reluctance to cut down whalecatcher fleets when others were unconstrained, but most of the companies found it worth while to join. The number of catchers in 1953/54 was 206 for 17 expeditions, compared with 230 for 16 expeditions in the previous seasons, a substantial economy in resources and costs on the catching side. Later efforts for agreements were less successful until the squeeze on profits intensified in the later years of the decade. One difficulty was that the interests of companies in cutting back catchers varied. The newest expeditions, with high processing capacity, needed a large catcher fleet to feed it, and wanted to prove themselves. Those with large by-product plants, like *Balaena* and our own expeditions, calculated the value of production lost from the reduction of one catcher a good deal higher than the Norwegians, who were still only producing whale oil. Their expeditions, more economical in fuel, could also see prospects of reducing, or even eliminating, tanker service by mustering a smaller catcher fleet. Sometimes those differing factors upset the limitation agreements, though the participants were well aware of the benefits to everyone of a smaller total fleet.

The industry's efforts to make such agreements were sharpened by the drop in whale oil prices from their postwar plateau. I was on *Venturer* in the Antarctic with Harold in early 1952 when a code telegram came in from our partner running sales, Tom Humphreys, reporting that the price had slumped from £100 to £75. That fall, which was permanent, signalled that the days of easy profits and generous spending were over. An immediate casualty was the little Labrador station, which had just survived on the high price but now had to be closed down. It was later sold to one of our ex-gunners, Johan Borgen, who struggled with it for a few seasons until it conveniently burned down.

To a whaler in 1954, pausing for breath and to take stock, the first nine years after the war seemed to have turned out reasonably well. The whale stocks were holding up and the price of whale oil had been high for most of the period. Awkward new entrants to the industry, the Russians and Dutch, were proving less aggressive than they might have been. There was one pirate, Onassis, but at least the

Whaling Agreements had not fallen apart. The Japanese, who even then looked likely to be the long-term winners in the industry, were expanding only gradually. We in Salvesen had a good fleet, steadily improving, and were consistently near the top of the season's performance tables. But the apparent stability and success was soon to be challenged. The second nine years marked the decline and eventual extinction of Antarctic whaling.

Five: 1954–1959

I went down to the whaling grounds for the third time in 1955. On the previous two occasions when I had been with Harold we had gone via Rio. On this occasion I flew further south to Buenos Aires to catch the Argentina de Pesca tanker, *Conquistador*, which was to take fuel oil to Grytviken. As might have been expected, *Conquistador* was delayed in her departure, so I had a full week in Buenos Aires. It was shortly after the fall of Peron and there was much excitement in the air, though having no Spanish at the time I could not take full advantage of my stay. Our company relationship with Argentina was uncertain. In 1949 one of our Antarctic ships had called there from South Georgia and been charged a hefty tax on the grounds that she had been carrying cargo under Argentine jurisdiction. Our agents paid up, to the great annoyance of Harold, who refused to reimburse them. After that we ensured that any of the ships that needed to make port in that area avoided Buenos Aires and put in to Montevideo in Uruguay, on the other side of the Plate. I did make a call on our agents, who, regardless of the past, were very helpful in telling me about the city. To find out their address I enquired at the British Embassy, where it was still the custom for visiting Britons to register. The lady in the embassy was firm. No, the embassy was not allowed to provide commercial information. I retired, baffled, but had the idea of asking the commissionaire, clearly a retired army warrant officer and likely to be more sympathetic. Yes, he said, their office was a few hundred yards down the street, just beyond the brothel. I couldn't miss it.

I was handsomely entertained by Alfredo Ryan, the owner of Pesca, at his ranch outside Buenos Aires. Ryan had started life as a shipyard worker in Gibraltar, a 'rock scorpion', and had emigrated to Argentina, where he had set up a successful ship-repairing business. As well as owning the whaling company he had interests in racehorse breeding and vineyards. The dozen bottles of his best vintage which he put aboard *Conquistador* for me were a considerable consolation on the slow voyage south.

Apart from tea at Harrods to the accompaniment of a string trio, and a bathe in the shallow and muddy waters of the Plate, I spent my time on whaling figures. Our whaling operations were tightly run, but our overall accounting was very primitive. We had no yearly budgets, and only the most hazy idea of the levels of catch and oil prices which the operations needed to cover their costs. This degree of ignorance was, I think, quite usual among industrial companies at the time. Harold, curiously for an economist, showed very little interest in that side. But the winds of change were blowing in industry. I myself was dimly starting to see that management accounting was an essential tool of efficient management at all levels. Having now emerged from Harold's tutelage I wanted to look at the whole economic basis of our operations. I was particularly interested to find out how we compared with the Norwegian companies in costs and whether our arrangements for repairing the catcher fleet over the winter at South Georgia were more expensive than bringing them home for repairs, as the Norwegians did. So I worked at this in the hotel with copious papers which I had brought down for the purpose. I came to some interesting conclusions, and wrote them up, to be pursued when I got home after the season.

While exploring the streets of Buenos Aires I found in a bookshop a treatise in Spanish on Argentina's claim to the Falklands. I thought it would be a good addition to the governor's library and bought it, later sending it to Port Stanley from South Georgia. I never heard whether it arrived safely or if he read it. I also tried to buy a slide rule there but my ignorance of Spanish foiled me. I should have been more persistent. When I did acquire a slide rule, six months later, it immediately became indispensable for the simple arithmetical calculations of yields, ratios, returns on capital, profit projections, which I now realised were needed to underpin most management decisions. Other buys in Buenos Aires were a leather cap, which I had admired on the heads of Pesca whalers (it turned out too small), and a tie to present to Arvid Rynning.

Rynning was one of the Leith Harbour originals. He started life with Salvesen as a chemist in the thirties. At that time he was renowned for the parties he threw in Tønsberg at the end of the season. He would invite all his friends, including some of the leading shipowners, to Hotel Klubben, where the best in food and drink was lavishly provided. After a few days his money from the whaling season would run out and he would report for a painting job on the factories at Husvik to keep going for the rest of the summer. About 1951 he decided to change his way of life. He settled permanently at Leith Harbour, taking winter after winter for seven years. He dropped drink and scarcely spent a krone in the slop chest. When he eventually came home he had collected a modest fortune. He made a happy marriage with the widow of Anders Herlofsen, one of the leading Salvesen gunners, and built up a prosperous business of his own. It is said that on his first visit to Leith Harbour Rynning was told by Leganger Hansen that there was a lighthouse at Cape Saunders occupied by a lighthouse keeper and two beautiful daughters. Next Sunday morning Rynning could be seen at the other side of the harbour, dressed in the best suit of a Tønsberg man-about-town, making his way among the tussock clumps and sea elephants towards the lighthouse, in fact a single flare supplied by gas bottles.

Conquistador was another ancient ship, barely able to do eight knots. With some bad weather on the way it was a fortnight before we reached South Georgia and were welcomed in at Grytviken. The enthusiasm of the reception committee that came round from Leith was, of course, not for me, but for the mail that the ship brought with her. The arrival of mail was more important for the whaler than anything else. Everyone's nerves went on edge as they waited for the sacks to be sorted and distributed. It was as if, having put it to the back of their minds for months, they suddenly remembered that they had a different life as well, of women and children and warmth and the smell of grass, and all their suppressed emotions welled up.

Leith Harbour had by then got through the main capital projects to modernise it. The new dry dock and the vast platers' workshop that we set up beside it at Stromness were completed. We had a big new tank farm to store whale oil and fuel oil for the three expeditions. Some of the tanks had been dismantled by the winter crews from the old Unilever station at Prins Olav and floated down the coast to Leith, a considerable feat of seamanship and rigging skill. The main addition on the production side had been a solubles plant, which took all the glue water from the cookers and evaporated it down to a viscous treacle. We had hoped that this new product could be shipped home in drums or tanks and sold as a high protein food supplement for cattle, but it didn't work out that way. The solubles were still 50% water, so they were awkward to incorporate in dry feed, and the amino acids which made up the protein were unbalanced. So instead we had to mix as much as possible of the solubles into the meat and bone meal as it was produced on the station. Eventually only a small tonnage was left each year to be sold on the market. With the recovery of all the glue water in that way very little now went into the sea, though there was still the blood and detritus from the cutting of the whales which drained off the plan to redden the water and excite the hordes of cape pigeons who lived on the morsels.

The station itself was getting more civilised. A cold store had been installed to expand the amount of frozen food that could be held. Our new tanker, *Southern Satellite*, had extra storage to bring down frozen supplies for the station. There was a large new cinema about to be built, and a more spacious hospital had been provided, an amalgam of the divergent views of several generations of whaling doctors. The railways, with their bogies propelled by muscle power, were being replaced by tractors and trailers. The biggest change from previous years was that there were fewer people. In 1951 there had been 600 men on the two stations, many of them tradesmen working on the new capital projects. As the projects were completed the tradesmen departed, but we continued to carry what seemed to the casual eye a very lavish establishment. Spurred on by the drop in whale oil prices we set about bringing numbers down. The station was still handsomely manned in 1955, but it at least felt like a whaling station rather than a huge engineering yard in which whaling was a minor activity.

I was only a few days on the island before sailing with *Southern Opal* to *Southern Harvester. Southern Opal* took us back to Leith later and again out to *Southern Harvester*. My companions were Ken Riddy, who looked after the helicopter operation, and Harry Kennedy, the ebullient catering superintendent, fizzing with ideas to transform our feeding. We enjoyed the company of Captain Baikie, a tough and outspoken master, who fortunately played bridge no better than the rest of us. Baikie kept a disciplined and clean ship, not always to be found in the general trading sections of the merchant navy. He was allergic to engineers and was barely on speaking terms with his chief. With our other tanker masters he provided an invaluable service to the whaling fleets. Their grouse, expounded by Baikie during my trips with him, was that although their crews shared many of the same hazards as the whalers they were paid much less on the agreed National Maritime Board merchant navy scales. There was not much we could do about that. Whaling rates were devised for men employed for a maximum of seven months. NMB rates, negotiated nationally for tramps and tankers, assumed year-round employment, and bonus arrangements were not appropriate for men who were spending so little of their service time in direct contact with whaling operations. Baikie was not one to give this up lightly, but he remained a disappointed man.

We had a more serious problem with pay scales in our tankers. For many years after the war good engineers were very difficult to come by, diesel engineers in particular. The best ones were taken by the passenger and cargo liner companies, the aristocrats of the merchant navy, leaving little of value for the rest of us. The National Maritime Board rates prescribed both minima and maxima, so we could not pay an engineer a premium to join us without breaking the NMB agreement. We had built up the tanker fleet from war-built steam ships, all we could get at the time, but there was no doubt that diesel was the future. This caused us much anguish. We bought our first diesel tanker, *Southern Atlantic*, in 1949, and were appalled to have to send it to sea with engineers who were alcoholic or manifestly incompetent. The easiest solution would have been to ignore the NMB agreement and pay what was needed to attract good engineers. But Harold would not contemplate that and we struggled to make do with what we could get on NMB terms. Eventually I did pay more, and hoped that he would not find out. Honour could be priced too highly. We also, in conjunction with other shipowners similarly placed, started our own training scheme. We took in young men as engineering cadets, trained them at technical college, and sent them to sea, bringing them ashore at intervals to take higher certificates of competency.

If our ships had been under the Norwegian flag there would have been no problem. Norwegian engineers started at the bottom as engine-room boys and worked their way up, coming ashore for a year at a time to study for their certificates at their own cost. In Britain the traditional recruitment for engineer officers was from time-served tradesmen in the shipyards. Ratings in the engine room had no ladder of promotion and engineers taking the 2nd or chief's certificates expected sponsorship from a company for their costs. Harold got the Leith Nautical College to launch a scheme for training up firemen to be engineers, as in Norway, and enlisted the support of his friends in the Seamen's Union. But the traditions of our merchant navy were too strong. Not enough young men of ambition were ready to take that route and it never caught on.

The 1955/56 season turned out to be an *annus mirabilis* for Salvesen. At the IWC meeting that summer the total quota had been dropped by 500 blue whale units to 15,000, in response to the worries of the Scientific Committee on the whale stocks. To sweeten the pill for the whalers the Commission abolished the sanctuary, set up in 1938 in the Bellingshausen Sea, west of the Antarctic peninsula, leaving the expeditions free to range over the whole of the Antarctic. The whalers had campaigned for this for some time. They pointed out that opening the sanctuary would not cause further depletion of the whale stocks, since that was governed by the total quota. The extension of the whaling grounds would take some of the strain off the present catching areas. Mackintosh and the Scientific Committee fought to defend the sanctuary. They argued that a collapse of the whale stocks might take place, so drastic that the remaining whales might be unable to reproduce themselves or even lose the will to survive. With such prospects it was essential to have an area where the whales continued to breed undisturbed. Harold poured scorn on this thesis. He maintained that commercial whaling would stop long before whales were reduced to that degree of scarcity. A future in which, in his words, 'boy doesn't meet girl' seemed to him total fantasy.

The main reason why the sanctuary had been readily approved in 1938 was that no one believed there were any whales to speak of there. In the 1920s floating factories had regularly fished from the harbours of the Antarctic peninsula which formed the eastern boundary of the sanctuary, but when Lars Christensen's expeditions had traversed the area in the early 1930s they had seen very few whales, so although the whalers welcomed the abolition of the sanctuary they were not prepared to try their luck there. Harold, contrary as always, took a different view. With the weight of his vast knowledge of the Antarctic to support him he could see no reason why there should not be stocks of whales there, and in reasonable abundance through having been left undisturbed for a generation. So he sent *Harvester* to fish there.

The results were spectacular. There were whales everywhere, and they had none of the craft to escape the whalecatchers which they had developed in other areas of the Antarctic. Helped by uniformly fine weather *Harvester* achieved 154,000 barrels in the fifty-eight days the season lasted, an average of 2,600 barrels per day. Since the maximum daily production of the factory was about 3,000 barrels she was virtually at full cook throughout the season. Most days the catchers were rationed and the full catch was taken before breakfast. The catch itself was not a record. In the early postwar years our own and other expeditions had gone well beyond it. But production per day in the very short season was higher than anything we achieved before or after.

I went out to *Harvester* by *Southern Opal* from Leith early in the season. The manager, Captain Bartho, was a conscientious master who had previously skippered a corvette tow boat. He was also an active member of Tønsberg Klubben, meeting place of the local shipowners and professional men, and a compulsive gossip. In the relaxed atmosphere of a season when everything was going right I used to stand with him on the bridge in the evening. In the intervals of guiding the movements of the fleet he would regale me with scurrilous stories of Tønsberg social and business life, sex and finance being intertwined as in the best Ibsen plays. His standard lead-in to the most spicy bits was 'Malicious tongues say' ('Onde tunger påstår'), a formula that relieved him from responsibility for the truth of his stories.

After a fortnight or so on *Harvester* the three of us moved, again by *Southern Opal*, back to Leith and then to *Venturer*, fishing the traditional grounds of the Weddell Sea. *Venturer* was doing quite well, but nothing compared to *Harvester*. Like everyone else I could not believe that the glorious catching with which *Harvester* had started the season would continue. But it did, and as the weekly catch figures came in to *Venturer* the manager, Captain Myhre, decided that he must move over and join *Harvester*. So for the first time in modern whaling history we sailed through the South Shetlands, past Deception Island, home to previous generations of whalers, guarded by the pinnacle rocks of Neptune's Bellows, and close to Smith Island and Elephant Island, where Shackleton's crew found refuge after the shipwreck of *Endeavour* in 1915. I took the opportunity of a helicopter flight as we passed through and had a panoramic view of the snowy rocks and seal-crowded beaches of the islands. When we arrived in the Bellingshausen Sea we were able to share in the catch abundance without disturbing *Harvester*. The weather helped, day after day of bright sunshine and calm seas, with the catcher decks hardly getting wet. Out on *Southern Lotus*, the leading gunner's whalecatcher, I was summoned to the gun platform by Arne Mikalsen and told to shoot the fin whale he was chasing. The boats were on ration and there was plenty of time to fill his complement. Mikalsen steered the ship so that the whale came up directly under the bow. All I had to do was to swing the gun, aim at the whale's emerging back and fire. Fortunately I did not disgrace myself. I got '*fast fisk*', though I had the usual fault of the novice in not firing quickly enough. I hit the whale in the back instead of forward in the heart, and it had to be reeled in to be given a death shot by a second harpoon.

The movements of the tankers during the season had been passed to my management, and I carried with me on my journeys the same battery of graphs and coloured crayons that I had previously maintained for Harold. As the season progressed it seemed quite likely that one or both of the factories would run out of storage space for whale oil. On reasonable projections of catch they had space enough, but if the exceptional catch levels of the Bellingshausen persisted and the stop date for the season, which depended on how quickly the whole Antarctic quota was taken up, was relatively late, we could be in trouble. So another tanker call might be needed to take away whale oil. The assessment was complicated by some of the figures being reported by *Harvester* which didn't make sense, showing that the stocks of whale oil on board were growing faster than the catch indicated. (It later emerged that Bartho was deliberately under-reporting his catch in the hope of discouraging *Venturer* from coming over to join *Harvester*.) I thought the risks were too great to leave to chance, so I ordered out *Southern Garden*, lying at Leith Harbour with a scratch crew loading meat meal from the station to take home. The manager, Bjelland, made up the crew from station workers without fuss, and off she set. *Garden* was not a brisk ship at the best of times, and her top speed of ten knots came down to eight or less when there was any weather. As she passed Cape Saunders she sailed into a full storm, where she floundered without progress for a week. By the time the storm had passed the crisis was over. The catching of the two factories had eased and an early end to the season had been signalled. So *Garden* turned back to Leith with nothing achieved. There was one benefit from the exercise. Thirty-five men or so had been taken from Leith Harbour to man her without apparently any effect at all on the station's work.

So we were strengthened in our determination to get numbers down to a lower level.

Harsh words were said to Bartho on his return home for his guile in concealing operational figures. But his star could not be other than high after such a splendid season. He continued as a successful manager for three further seasons. Unfortunately in 1958/59 he got at odds with Hans Abrahamsen, the leading gunner of *Harvester*, by taking the factory ship in one direction when Abrahamsen had decided that the whales were to be found in another. There were a few whaling managers sufficiently strong and experienced to override successfully the views of their gunners, but Bartho had not quite achieved that position. Indeed any manager would be unwise to set himself against Hans Abrahamsen, for thirty years one of the Antarctic's outstanding gunners. So with regret we parted company.

I went home at the end of the season by *Southern Lily*, one of our converted corvette catchers, which dropped us at Rio, grown to be almost unrecognisable in comparison with my first visit seven years before. I might have sailed back with one of the factories, but that would have taken three weeks longer and there was plenty to do at home. On getting back to the Leith office I produced with some pride the conclusion of my laborious calculation in Buenos Aires that our profit performance was markedly poorer than our Norwegian competitors and that repairing the catchers at South Georgia was an extremely expensive operation. This was not received with any resounding appreciation. Harold told me to go away and do something about it. But we did get down seriously to analysing and costing our whaling operations. Bernard Street that summer was overrun with calculating machines and sheets of pencilled figures. At the end of it we had satisfactory cost figures for the two floating expeditions and Leith Harbour. From then onward we could make satisfactory budgets for each season.

There had up to then been a curious separation between the operations of the company and our yearly accounts. It is normal in businesses for each activity to be surveyed throughout the year in profit terms and for them all to be brought together at the end of the year in a profit and loss account. Not so with Salvesen. On the whaling side none of us knew what the year's outcome was until the annual accounts were completed nine months after its end, though the Chief Accountant might just be persuaded to make a cautious guess before the figures came out. For the merchant ships voyage accounts were produced promptly but no repair costs were charged. Harold who was, rightly, sceptical about the ability of our merchant fleet to earn money, said this practice was devised to make actual losses look like profits.

The reasons for this detached attitude to annual accounts lay in the nature of the company. Harold was interested in profits, and successfully made them for the company, not only in whaling. But he was not fussed about fluctuations of profit from year to year. The company's businesses were whaling and shipping. Provided they remained sound it did not matter if during the inevitable cycles there were years of loss. He saw it as his duty to build up the asset base of the company. His generation had received it in this trust from their fathers and should pass it on to their sons. He did not like paying out dividends, which he thought an unjustified indulgence, and was deaf to the suggestions by some shareholders that they might be allowed special payments in good years. The family had inherited the shares and should be content to pass them on, made more valuable by what had accrued in the company during their lifetime. One way to preserve wealth was to pay as little tax as possible. For many years shipping investments were allowed free depreciation, i.e. the whole cost of a ship could be set immediately against profits, so reducing, even eliminating, payment of tax. That also kept at bay the department of the Inland Revenue which watched 'close companies' (i.e. those controlled by a few shareholders) to make sure that they did not keep profits in the companies instead of paying them out to shareholders, who could then be handsomely taxed. If the profits were being invested in the business the Inland Revenue did not insist on high payouts.

For Harold the ideal shape of an annual account was an initial profit turned into a loss by charging against it the whole cost of a new ship. He could then resist any pressures from shareholders or the Inland Revenue for a high dividend payout while continuing to build up the value of the company.

Harold's practice did not entirely follow these austere principles. The company did pay a steady though low dividend, based on the pessimistic assumption that there would be no trading profits and that it had to be covered by the income from the reserve investments in equity shares. The company's capital was accumulated, but not hoarded. It was always available to be invested at risk for the company's development, and Harold, like his predecessors, was an entrepreneur. Nor did he allow his loyalty to the Salvesen family, which was intense, to get in the way of recruiting managers for the company from outside. By setting almost impossible standards of entry he frustrated shareholders who had the traditional belief that a family-owned company should also find business careers for family members.

Profit performance of our whaling in the postwar years did fluctuate considerably with levels of catch and whale oil prices. In the early seasons when food oils were scarce and catches high we did very well. The first season with both *Venturer* and *Harvester* working, 1946/47, gave us £3.6m. before fleet depreciation, enough to pay back completely the cost of the two new floating factories and much of the new catcher fleet. The following seasons gave less spectacular results, but, apart from a poor year in 1952/53, profits before depreciation averaged well above £1m. until after 1956. They then moved to a much lower level, declining towards a bare break-even. Though there was a steady capital expenditure to improve the fleet and equipment, these surpluses allowed us to increase our reserves as well as financing the new fish factory ships and some merchant ships. By 1960 our reserves, invested in equities, had gone up to £13m. They continued to grow after the end of whaling, and were of great benefit to the company when we started to invest on a large scale in new activities.

I used to study the accounts of our best Norwegian competitors as they came out. They generally looked better than ours, although their catches were less. Part of the difference was that with their all-Norwegian crews and their single base in Norway their costs of administration and repairs were a good deal lower. It seemed evident that bringing the catchers home to be overhauled by their own crews was cheaper and more effective than running a repair yard at South Georgia. Their floating factories, larger and with diesel motors, generally needed less tanker service than ours. Then there was the complication of by-products, which we had and they did not, principally meat meal. Our income from meat meal was significant, but it may well have been swallowed up by the extra crews required, the high costs of operating and repairing the plants and the loss of tank space. Several of the Norwegian companies decided, in the early 1950s, to lengthen their factory ships by cutting them in two and putting in an extra section midships. That gave them several thousand tons more of tank space and relieved their need for supply by expensive tankers in the Antarctic. We thought of doing this but decided against it. It would have meant an entire reconstruction of the complicated factory plants, and there was the risk that some hold-up in the shipyard might find one of the floating factories still in two pieces when she should have been on her way south for the season.

We managed over several seasons to bring down the production manning required on the floating factories. We also streamlined the organisation for catcher repairs at South Georgia, not only the winter crews themselves, but the extra manning at the station during the season attributable to maintaining the repair base. So costs came down, though operational savings were limited by the design and size of the floating factories and tanker fleet. Some of our higher costs were in special equipment like asdic or the helicopters. They should have justified themselves in higher catch, though it was difficult to calculate convincingly what they contributed. If we had had the opportunity to build an entirely new fleet in, say, 1955, we would have had a much bigger floating factory, of at least the size of the South African *Abraham Larsen*, ex-*Unitas*, with 30,000 tons deadweight capacity and diesel engines. The catcher fleet would have been of 17 knot diesel boats. With that combination the expedition could have made and stored a full season's oil and by-products and been self-sufficient in diesel oil without any supporting tanker. But new investment to that degree in a declining industry was unthinkable. Only the Dutch and the Russians could do this. The Dutch company, still determined to make a go of it against all the odds, and backed by state aid, replaced their small floating factory with a new large *Willem Barendsz* in 1955. The USSR decided on a large building programme about the same time, the first new factory expedition being intended to reach the Antarctic in 1958.

The catch figures of the two Salvesen floating factories in 1955/56 were the best since 1948/49, but they stood out against a clear downward trend. The total quota had been brought down by the International Whaling Commission (IWC) from the original 16,000 blue whale units to 14,500 in 1957/58, so there was 10% less catch available for each expedition. The composition of the industry was also changing to the disadvantage of the traditional operators. Up to the middle 1950s the Norwegians and British still preserved their dominating role in the Antarctic, with thirteen out of the nineteen floating factory expeditions operating in the 1955/56 season. The

Japanese were however now starting to flex their muscles; having been cautious in re-establishing themselves in the industry after the war, they now saw good opportunities to expand.

The older operators, particularly Norway, tried through diplomatic pressure to restrain Japan. The Norwegian whaling companies and their government pointed out that the entry of more Japanese expeditions would do nothing to increase the world's supply of whale oil and would simply drive out other operators. Norway had limited the rebuilding of its fleet after the war for the benefit of the whole whaling industry and to preserve whale stocks, and now pressed others to follow the same path. This approach had sympathy, but not support, from the US and British governments, who felt that an attempt to impose commercial restrictions on Japan might provoke an anti-Western reaction. The Japanese government was in any case not impressed. Whaling was a traditional Japanese industry and whale meat had always been an important part of the national diet. If there was virtue in full utilisation of the whale the Japanese whalers could claim recovery far beyond the Europeans, whose interests were largely limited to oil and meat meal. So expansion of Japanese expeditions went ahead, increasing from three in 1954/55 to five in 1956/57, and steadily progressing thereafter. But the Japanese government, which by its licensing powers controlled the policy of its companies, did pay some heed to the arguments about excessive capacity in the industry and pressure on whale stocks. After 1956 additions to the Japanese fleet were authorised only through purchase of existing whaling expeditions. This eventually moved virtually the whole European industry, except for USSR, into the hands of the Japanese.

The Japanese companies were formidably efficient operators, as they had already shown before the war. The Norwegians and British themselves had most of the right ingredients, but the Japanese had some additional advantages. Because of their interest in whale meat and even blubber for food they had a strong home market for highly valued products. Their crews, lower paid than ours, worked with the enthusiasm and discipline of a national army. In catching power they improved each season, with heavy investment in large diesel catchers, seventeen knots being the norm against our fifteen. They had been independent of Norwegian gunners almost from the beginning, and their gunners were at least as good as ours – probably better – because younger, as was to be expected in an expanding industry. Even in 1955/56, when *Southern Harvester* created an all-time record in catch per day for any British or Norwegian expedition, it was topped by the Japanese *Nishin Maru*, without the advantage of operating in the 'sanctuary'. With such success the Japanese companies were able to pay good prices for European expeditions even when the end of whaling was in sight.

Olympic Challenger was the first expedition to be sold to Japan. She had operated since 1950/51, with one season away when it was more profitable to Onassis to work her in the tanker business. From the beginning she had ignored the rules of the International Whaling Commission, catching outside the prescribed season and taking whales indiscriminately of any size or type. This soon became well known to the rest of the whaling industry and their governments. There was no quick way of stopping such piracy. Panama, which as a member of the IWC should have taken responsibility for *Olympic Challenger*, simply repeated the outright denials of Onassis. It took time to collect irrefutable proofs. Some of them were produced by other operators who came across Olympic catchers on the whaling grounds. Others were obtained from information collected from crew members, mainly German, who felt that the truth should come out. The Norwegian Association of Whaling Companies was active in compiling a dossier. Eventually a public complaint on infringements was made by Japan at the 1955 meeting of the IWC, backed by strong evidence, and the matter was in the open.

Meanwhile *Olympic Challenger* had provoked international conflict in a different area. In the late summer of 1954 the expedition set out to catch sperm whales off the Peruvian coast on the way to the Antarctic. Salvesen floating factories had fished there in the late 1930s and *Venturer* had a brief season in 1947. But the political position had changed. Peru and Chile, which had refused to join the IWC, had declared a 200-mile limit of sovereignty over the Pacific Ocean adjacent to their coasts. This was the first step in the establishment of wide fishing limits by coastal states to protect what they considered their national resources. The 200-mile limit has since become an accepted norm, but at the time it seemed a gross infringement of the traditional freedom of the seas and met strong protest. Onassis, whose respect for international law, past or future, was slight, decided to try his luck, with a precautionary insurance at Lloyd's against the remote prospect of interference by the Peruvian government.

USA and Norway, as leading members of the IWC, had conflicting loyalties here. On the one hand they had to oppose vigorously any move to shut off international access to high sea fisheries and possible restriction on the movement of merchant ships; on the other they could hardly put their support behind a whaling crook. *Olympic Challenger* was not only breaking the provision that stopped Antarctic expeditions from fishing for baleen whales outside the Antarctic, she was also likely to be observing no size restrictions, playing havoc with the small sperm whales of the Humboldt current. (As indeed she did. When her correct catch figures were revealed it turned out that 95% were under the IWC size limit.) The Peruvian government had given warning that it would enforce the 200-mile limit. In November 1954 it sent out two destroyers and a plane to arrest the factory and bring it in to the northern port of Paita. The leading gunner, Johan Borgen, an old Salvesen gunner, wanted to fight, and offered to sink the Peruvian destroyers by harpoon, but the German captain decided to go quietly. The expedition was fined $3m. by the Peruvian courts. She paid up, being reimbursed by Lloyd's, and was released just in time to get to the Antarctic for the 1955/56 season.

That season was the last for *Olympic Challenger*. Soon after her return the Norwegian companies released in full detail the information they had collected in sworn depositions from German crew members, photographs of undersized whales on the *Olympic Challenger*'s plan deck and logbook entries. Opinion in Germany, where the expedition was based, did not entirely welcome these revelations, for Onassis was a good customer of German shipyards and had revived German whaling ambitions. But the Norwegians were able to persuade the local court in Hamburg to arrest *Olympic Challenger* in anticipation of a Norwegian law suit. Onassis retaliated by getting Jahre's *Kosmos III* arrested. He had by now sold *Olympic Challenger* to Japan, and needed to hand the ship over in good time for the next season, so he was keen to get a quick settlement. He closed the matter by passing a substantial sum to the Norwegians. There was a face-saving formula attached to it but, in the eyes of Norway and the rest of the whaling world, some justice had been done and the principles of international conservation upheld.

To the economic assessments I had brought back from the Antarctic in 1956, which had cleared my own thinking, if not anyone else's, I had added an epilogue to the effect that it looked as if

Antarctic whaling for us had no more than seven years' life left and that it would be sensible to sell one of the expeditions now to get the benefit of the high prices being offered by the Japanese. It was easier to put that on paper than to face the reality. The lives of all of us were bound up in whaling and the prospect of dismantling that great enterprise was terribly upsetting. But led by Harold, even more emotionally involved than the rest of us, we accepted that we had to act commercially. In 1956 we were approached by a Japanese company with an offer to buy *Southern Venturer*. Japan had just bought *Abraham Larsen*, the massive ex-German *Unitas*, which we had coveted after the war, but remained keen to expand. However the Japanese government, no doubt sensitive to possible reaction from the older whaling countries, decided to limit transfer to one expedition a year, and refused a licence. Later the Japanese turned their purchase interest elsewhere. It was not until 1962 that *Venturer* was sold.

My prediction in 1956 of seven years more whaling for Salvesen turned out to be exactly correct, but it was a pure guess. We still knew very little about the Antarctic whale stocks beyond the evidence we had from our catching experience that they were declining. In any case that might not turn out to be the most important factor. The arrival of new competition, the development of the IWC régime, changes in the price of whale oil, better utilisation of the whale, would all contribute to determining our remaining span of life. There was no firm basis for making a sensible forecast, and any prediction we made was likely to be quite wrong. All we could do was to carry on from year to year, investing to keep our operations efficient, and hoping that we could combine some benefit from Japanese buyers with a gradual run down. In the event that is what happened.

The *Olympic Challenger* interlude distracted attention from the equally serious flouting of the whaling convention by the Russian expedition *Slava*. Expeditions who had operated close to her in the Antarctic continually reported having observed *Slava* catchers fishing outside the season or taking whales which were protected. Evidence of illegal catching was sent in to the IWC, which had a special committee appointed to review infractions. The USSR representatives of course totally denied any illegal catching, claiming that the sightings were wrong or even imaginary. The other governments represented in the IWC were not naive; they realised that the

Russians were cheating, perhaps on a large scale, but it did not seem fruitful to pursue them. Public accusation might have driven the USSR to resign from the IWC and slip completely outside international control. Japan was particularly wary of criticising the Russians. Their chief delegate to the IWC, Fujita, spent most of the year in Moscow negotiating on Pacific salmon rights, and he had no wish to offend Ishkov, the powerful and long-serving fisheries minister of the USSR. The whaling companies did not take such a diplomatic view. They pointed out that reductions of the whale quota were imposing sacrifices, even bankruptcy, on them while the Russians were untouched. They were asked to reduce their whaling, not for the benefit of the whale stocks, but to enable the Russians to take more. The operation of the existing régime did not only allow the Russians to take what they wanted; they could also, and apparently did, reduce the amount the rest of the fleet took by declaring figures far in excess of what *Slava* actually caught. By reporting each week an exaggerated number of blue whale units as their expedition's catch they inflated the total catch figure and ensured that the blue whale unit limit was reached earlier than it should have been, so the other expeditions were deprived of catch. As the pressure on the whaling companies grew with the reduction of the overall quota their bitterness against the Russians increased, together with their conviction that neither the whale stocks nor their own operations could be saved unless the current IWC catching régime was scrapped.

The Russians were unashamed about their behaviour, and treated the questions and denials at the IWC as a formal charade. On one occasion I met some of the representatives of their fisheries ministry in Newcastle where they had come to inspect *Venturer* as a possible alternative to one of their planned new buildings. As we sat in the manager's cabin drinking coffee after their inspection I asked them why they went on cheating. They had to, they said, until they had built up their fleet to parity with the West. Then they would stop.

The Japanese expeditions were also none too scrupulous in their observance of minimum sizes, though they were never found to be taking protected species or operating outside the season as the Russians were. There was a revealing feature in their catch of pygmy blue whales, a stock which they discovered on eastern grounds well north of the normal whaling latitudes. These whales seemed to have a length at maturity well under the normal blue whale, and quite close to the minimum permitted size length. The reports of their catch showed a heavy cluster just above the minimum length, indicating that a number of undersized whales had been stretched to make them legal. Norwegian expeditions also reported that they had fished for sperm in the same areas as Japanese expeditions and had had to give up because most of the whales spotted were undersized. The Japanese had carried on catching and the whales had appeared in their reports as over the minimum.

The dismay of the whaling companies over Russian behaviour became more acute when it became clear that the USSR had embarked on a large new-building programme. At least three floating factories were under construction, two of them the biggest whaling ships ever, and they were to be served by 17 knot catchers of 3,600 h.p. Faced with this competition, the British and Norwegians could see no future for their own expeditions, confined within the Convention and bound by catching rules which the Russians ignored.

It now seemed clear that the only possible salvation for the Western whaling fleets and for the whale stocks was to scrap the régime of free competition within a fixed total quota and get back to separate national or expedition quotas, which had been quite successfully applied in the 1930s. Harold had been keen on this at the time of the setting up of the International Whaling Convention in 1946, but the Americans found it contrary to their ideas of free enterprise and the Norwegians were afraid of new countries coming in and claiming permanent property rights in the Antarctic catch. Now the British and Norwegians were of one mind. An allocation of national quotas would not stop the Russians from cheating, but it would remove the hold they had at present on the catch potential of the others through reporting bogus catch figures. If a country had allotted to it an agreed share of the catch it could rationalise its capacity, reducing the number of catchers and spreading catch over the season to get the best possible utilisation in factory production. As the total catch ceiling was reduced, as it would have to be to save the whale stocks, each country would send out fewer expeditions but could still manage to maintain profitable whaling.

The International Whaling Commission worked quite well in regulating whaling apart from the all-important issue of reducing the overall catch and preserving the whale stocks. It decided the minimum length of whales of each species allowed to be taken and pushed back the opening date of the season to later in the summer, when whales would be fatter and give better yield. It imposed separate limitations on blue whales and humpbacks when it appeared that they were being overtaxed. Blue whales were from the beginning preferred to fins because of their size, and stocks had already seemed to be declining in the 1930s. Humpbacks were very easy to catch and concentrated in a few areas, particularly south of Australia. Eventually, but only as late as 1965, blue whales were completely protected. Humpback catching, after some years of limitation to a small quota that was taken in a few days, was totally banned in 1963. These measures, as with all whale conservation decisions, came too late to stop the reduction of those species to a very low level, from which they are still only slowly recovering.

Decisions in the IWC were taken by majority vote, but there was no way of forcing countries who disagreed with a decision to conform to it. Any country had the right to lodge a protest against a decision and would then not be bound by it. That meant that effective and binding decisions could only be taken by consensus among those concerned. If one whaling country, having been in a minority on an IWC vote, protested, the others, whether or not they had voted for it, would have to protest as well so that they would not be subject to a restriction which did not apply to their competitors. So it was that in 1958 when Holland alone voted against the limitation of catch to 14,500 units, and protested, all the other whaling countries had to protest as well, and the catch ceiling reverted to 15,000.

Accession to the International Whaling Convention and so to membership of the IWC was open to any nation that chose to join. Initially the signatories were the Antarctic whaling countries and those who had interest in coastal whaling in other parts of the world. The USA, which had taken the lead in setting up the Convention, pressed consistently for conservation. The coastal whaling states, particularly Canada, Australia and New Zealand, argued strongly for lowering Antarctic catch ceilings and restricting pelagic whaling elsewhere, not forgetting that this would help their own land stations to survive. Of the Antarctic whaling countries, Britain and Norway, with their long-established interests in the Antarctic, were the most conscious of the decline of whale stocks, and tried to bring about a gradual reduction of the catch ceilings, following the recommendations of the IWC's Scientific Committee. But they became increasingly worried about the leakages in the IWC régime. Japan was less concerned. Her fleets were fishing and expanding. The Japanese government was ready to go along with catch quota reductions, but did not feel it necessary to press for them. Indeed Japan voted with Holland against a quota reduction in 1956, though it accepted the majority view. The USSR did not much care about what the IWC decided as long as, by fair means or foul, its newly expanding fleet was left free to catch what it could. The Russian scientists who attended IWC supported the forecasts of doom sent out by the Scientific Committee, while Solyanik, representing the industry, declared roundly that there was no shortage of whales. It seemed to us extraordinary that the planners in Moscow should have decided on an expansion programme involving at least a fivefold increase in the USSR catch, when their own scientists were reporting massive declines in the whale stocks.

The Dutch government took a position radically different from everyone else, with a policy apparently shaped entirely by the demands of national prestige. Holland had entered Antarctic whaling in 1945, picking up its whaling traditions of 300 years earlier. With the active support of the state and in the teeth of opposition from the established whaling countries, the Dutch had survived in the Antarctic, if not prospered. Its government policy was now focused on keeping the Dutch company in business regardless of economic results or the preservation of whale stocks. The services of their scientists were enlisted in support. Dr Slijper, the Dutch scientist on the Scientific Committee, consistently challenged the view held by all the other members, maintaining that there was not sufficient evidence to show that the fin whale stocks were in decline and that the quota needed to be reduced. He even went so far as to suggest, at the meeting in 1953, that fin whale stocks might have increased since the war. The considerable work done by British and Norwegian scientists, led by Dr Mackintosh and Professor Ruud, could only produce rough assessments of whale stocks, but the trends were clear enough and Dr Slijper's dismissal of their evidence put into question his status as a scientist. Resting on this, however, the Dutch government opposed reduction in the catch ceiling when the IWC voted in 1956 to bring it down from 15,000 to 14,500. On that occasion it went along with the reduction, but in 1958 its vote against 14,500, and subsequent protest, brought the quota back to 15,000 at a time when the Scientific Committee believed that a reduction to 11,000 units, or less, was needed to preserve the stock.

Harold Salvesen, sharing the role of industrial adviser to the

British delegation with Rupert Trouton of Hector Whaling, took an active part in the IWC negotiations, particularly as the pressures of stock decline, competition and cheating became more intense. Harold had no illusions about the decline of the whale stocks. As early as 1945 he had warned Salvesen shareholders that the whaling industry had a limited life. Without taking the initiative in proposing a reduction of catch, he accepted the general policy of the British government of getting catch quotas down to a sustainable level. He was particularly concerned about the cheating, and argued fiercely that there was no point in reducing catch quotas under the existing régime if the rules were ignored. But Harold, true to himself, did not swallow intellectually the generally accepted thesis that a self-renewing natural resource should only be harvested at a level that allowed it to be sustained in the future. He would point out to me that in economic terms it might be right to use the existing capital equipment to take as many whales as possible up to the point when there were so few left that they were not worth catching. That would give immediate benefit to the whaling companies and the world's supply of wealth. Whale stocks would be brought to a low level, but not so far as to be in danger of disappearing, and they would eventually recover. Against this had to be set, in world terms, the loss of the far greater wealth that the whales could bring in the future if harvested as a sustainable resource. But, he argued, even if we rate the interests of future generations as high as our own, it is by no means certain that whales will have any economic value in the future. Their products may well be replaced by others that can be produced at less cost. Harold never pursued this line of thought in public. He was as conscious as anyone of long-term world interest, but he felt that sustainable harvesting, like any other policy, should be argued out and not accepted as self-evident.

The IWC had provisions that the whaling countries should appoint two government inspectors to each of their floating factory expeditions who would ensure that the regulations of the Commission were strictly observed. The inspectors were required to report any infractions to their government, and these were scrutinised by the Infractions Committee of the IWC. On the British, Norwegian and Dutch expeditions the inspectors, as might be expected, did their jobs punctiliously. The only infractions that occurred were in undersized whales. There were always a few of those, when the gunner had misjudged the length, but it was a matter of professional honour to take only legal whales and crews in any case were paid no bonus on illegal ones. If a gunner became careless he was firmly pulled up by the manager.

The inspectors who sailed with our expeditions were men of quality. Some were ex-naval men, supplementing a modest pension with the equally modest inspector's salary. They became valued members of the ship's company, sometimes taking on the job of preparing the daily news bulletins which were captured by the radio officers, typed and distributed. Some were young scientists researching on whales, who could work on the carcasses as they came up the slipway and passed over the aft and fore plans.

The IWC provision for government inspectors was clearly inadequate against governments who ignored the rules, such as the USSR, or who, like Japan, were ready to overlook collusion between inspectors and their expeditions. The Commission had no authority to provide its own inspectors. The only solution seemed to be the appointment to each expedition of an observer from another country who would check that the national inspectors were doing their job properly. That was proposed by Norway at the IWC meeting of 1955. It would have been too complicated to amend the Whaling Convention to embrace such a scheme, so it was accepted that a separate agreement would have to be made between the five pelagic whaling countries. In subsequent years Norway and Britain pressed for such a scheme, and they soon had the support of all the whaling countries except the USSR. The Norwegians were initially keen on having an observer scheme without the Russians, hoping that they would come in later. The UK, on Harold's advice, opposed that. The observer scheme was intended to stop the Russians cheating, and without the Russians there was no point in it. Discussions on an observer scheme went on from year to year without conclusion. As will be seen, it was not till 1963 that a scheme was agreed between the whaling countries – and that was not the end.

By 1958 a turning point had been reached in the IWC. The Norwegians and British realised that if they were to have any chance of survival the free competition régime must be scrapped and replaced by a system of national quotas. When initial discussions were started the Japanese were lukewarm. They had little to lose in continuing the existing régime which was playing such havoc with their competitors. The Dutch, short-sighted as ever, wanted to get out of the Convention and fish without any catch ceiling. The Russian attitude was still unknown. Harold thought it would be impossible to get them to join a quota agreement, but that the other four countries should still be ready to act without them.

Negotiations started in the autumn of 1958. To everyone's surprise the USSR decided to take part and, after a brief haggle, agreed to accept a 20% share of the total catch ceiling as their quota. That figure was a good deal more than the catches they had been taking with *Slava*, and than those that they had been declaring. But it was less than the share they might have claimed after the addition of three large new expeditions to their fleet. This breakthrough gave new heart to the European whaling countries. They appreciated that the USSR would probably go on as before, taking as much as it could, but they now had the opportunity to negotiate national quotas which they could catch regardless of the Russians' behaviour. The way would also be open to reduce fleets to make the best economic use of each national quota and meet the impact of the progressive lowering of the total catch ceiling. Experience has shown that the only effective way to preserve whale and fish resources on the high seas, traditionally held to be common property, is to create national, company, or personal property rights in them. This was now happening in Antarctic whaling. The doctrine of free competition, enshrined by the USA in the International Whaling Convention of 1946, had failed.

If quotas of a fixed total catch were to be allocated nationally it would have been logical to take a further step. Antarctic whales could be considered as belonging to all the nations of the world, not just those who happened to have a stake in the whaling industry. On that premise a world authority, if there was one, could have claimed possession of the stocks. It might then auction the right to exploit them to the highest bidders and use the money it received for whatever purpose it thought fit. I developed that idea in a paper to the Scottish Economic Society in 1959 and John Gulland of FAO (Food and Agricultural Organisation of the United Nations) came up later with something similar. Harold dismissed it as a useless intellectual exercise. Such an arrangement could of course have had no practical application for the IWC and the 1958 quota agreement, but it may yet give some guidance in the management of national, or even international, fish stocks.

The 1958 agreement was to run for seven years. It specifically allotted 20% of the IWC global quota to the USSR, and the remaining 80% was to be divided among the other four whaling countries 'in a way satisfactory to them'. Expeditions could not be transferred between countries unless part of the national quota went with them or the factory ship was withdrawn from Antarctic whaling. Finally, with the spectre of Onassis still visible, it was provided that if an expedition under a new flag entered the Antarctic without being a member of the IWC and securing the transfer to it of part of one of the national quotas, the agreement would be null and void. So Antarctic whaling was transformed from an open game to a closed club, though one which could have been smashed open again at any time.

The Norwegian government had taken the lead in securing the agreement with the USSR. But that was only the first step. The other whaling countries now had to agree the allocation of the remaining 80%. If there was no agreement the IWC provisions would remain in force with free competition for 80% of the overall quota, which would have been intolerable. So at the end of 1958 Norway announced its resignation from the Commission, making it clear that the alternative to a national quota agreement was not continuation of the IWC régime, but uncontrolled catch, which would quickly destroy the whole industry. It was a grave step to take. Norway, as the biggest whaling country, had been the principal champion of the IWC and the most faithful to its ideals of rational conservation. The whaling statistics of the Antarctic were collected by an international bureau run by Norway. The whaling expeditions sent their weekly catch reports there, and the bureau decided from them the date when the overall catch ceiling was deemed to be reached and the season was terminated. Norway's departure seemed to imperil the already fragile frame of international cooperation in the Antarctic that had been so carefully built up. Norway was fully aware of what it was doing. The IWC in its present shape was no longer able to preserve either the whale stocks or their national industry. Norway genuinely expected to rejoin when a full quota agreement was signed, and such an agreement was intended to work within the IWC. Meanwhile the risk of a new anarchy had to be taken.

When Norway resigned from IWC there was no rush to follow suit by the other whaling countries. The only one to do so was Holland, which saw an opportunity to pursue its solipsist policy of free catching. Britain did not want to break its general policy of supporting international agreements. The British whaling companies would have preferred a resignation, both to support the national quota negotiations and to make sure that in the event of a breakdown the British whaling companies, still within the Convention, would not be unfavourably placed compared with the Norwegians and others outside. But they did not press the British government hard. With the removal of the Norwegian expeditions the overall catch ceiling at its existing level would no longer put any restriction on the catch of the remaining whaling countries.

It had been hoped that a full national quota agreement would have been signed in 1959 in time for the 1959/60 season and that Norway and Holland would have quickly come back into the fold, but that did not happen. As will be seen later it took three years more to bring a full quota agreement to completion. Meanwhile for 1959/60 each country declared a quota for its expeditions. The USSR was content with 20% of 15,000, the ceiling eventually set for the previous season. 3,000 units were quite enough to cover the catching capacity of *Slava* and the first of the new expeditions. The others set quotas which together took much more than the remaining 12,000 units. The total catch ceiling came up to 17,539 units.

Thus the result of the Norwegian and Dutch resignations from the Commission was a large increase in the potential catch. The ceiling for 1958/59 had been voted at 14,500 units by the Commission, but had had to be increased to 15,000 when the Dutch protested. It was now 20% higher than the IWC figure, and nearly 60% higher than the figure of 11,000 which the Scientific Committee had estimated as a maximum sustainable yield. That did not bode well for the future.

Immediate proof of the run down came with the catch figures of the 1959/60 season. Despite a longer season than 1958/59 none of the whaling countries, except Japan, managed to reach its declared quota. Against the 17,539 quota only 15,512 units were taken in a catching season of 102 days. Only four seasons earlier, in 1955/56, 15,000 units had been taken in 58 days.

The reversion to virtual free fishing in 1959 gave our own expeditions some short-term benefit. We were able to work with fewer whalecatchers, and were now insulated from the intense competition for a share of the overall catch quota. But the scarcity of whales was showing itself increasingly in our catch figures. In the great year of 1955/56 *Venturer* and *Harvester* had taken 280,000 barrels between them. In 1958/59 the total was only 135,000, and the catch at 157,000 was little higher in 1959/60, after the suspension of the IWC régime.

Meanwhile the whale oil price, up at £85 in 1955/56, had declined to little over £70 in 1958/59, and was to go further down. In 1957/58 and 1959/60 our whaling operations barely broke even before charging depreciation. In 1958/59 we recorded a £200,000 loss at the same level.

We might have felt like throwing in the towel at that point, but a new product kept us going. From time to time whaling factories had experimented with the production of meat extract, starting with the Germans before the Second World War. *Balaena* had always run a small extract plant. In the middle 1960s the market expanded and it became accepted as a substitute for beef extract in dried soups, indistinguishable in taste. Though there were some problems in EEC labelling laws, where Liebig fought hard and unscrupulously to preserve their monopoly in meat extract, the whale product sold well, and at the high price of £1,000 per ton. Encouraged by Rieber of Bergen, soup makers themselves, we put in a plant on *Harvester* in 1956/57. The early stages of production, cooking the meat and pressing out the liquids, were identical to those of a meal plant, so it was easy to add extract production machinery to our existing meat meal lines. The first year's production was only 67 tons; the process looked simple, but we found we had a lot to learn. Next year all three expeditions had extract plants, with a combined production of nearly 400 tons. In 1959/60 the total rose to 880 tons, and in 1960/61, the last year we operated the two floating factories and Leith Harbour, we reached over 1,300 tons. At that point we were by far the largest meat extract producers in the world. Meat extract, packed in steel barrels, was the ideal whaling product, combining high price with very little space demand. At Leith Harbour, where there was plenty of time to work up each whale thoroughly, we found we could take out nearly twice as much extract material for each ton of meat as on the factories, and figures of yield reached a very high level. Leith Harbour had however one particular problem: the whales were being taken further and further away from the station, with sometimes up to thirty hours' towing to bring them in. By that time the meat normally would start to decompose and become unsuitable for meat extract production. Old whales also gave low quality oil. We countered this by injections of antibiotic, and got the gunners to slit the carcasses so that the innards were exposed to the cold sea water during towing.

The happy state of unlimited demand at a high price could not last. Norwegian expeditions put in meat extract plants at the same time and by 1961 there was a serious overhang of unsold product. The Norwegians, with whom we were cooperating in a selling pool, pressed for production to be restricted in the 1961/62 season. I was for going on with full production, since the end of whaling was in sight and it seemed sensible to produce while we still could, storing any surplus until the market came back into balance. (We had found that whale extract, like port, improved with age.) Harold had taken a similar view in 1931/32, when the Norwegian fleet laid up and Salvesen and Unilever operated alone. But I got no backing from the Norwegian companies. We decided to support rather than fight them on that occasion, and closed down the *Venturer* plant in 1961/62, leaving only *Harvester* to produce extract. My instinct to go on producing was justified earlier than expected. The market recovered during the 1961/62 season and we were even able to arrange for *Venturer* to open up her plant and get half a season's production.

An alternative product to meat extract was frozen meat. *Balaena* had maintained meat frozen for pet food as a principal product, but none of the other European expeditions had touched it. Our painful experience with *Southern Raven* in the 1940s had kept us right out of that area. The frozen meat market became interesting again in the later 1950s. Petfoods, who had large requirements for cat and dog food, were hunting for new sources of supply and persuaded one of our competitors at South Georgia, Husvik, to put up a plant. It was not an economic success, since the whales, as noted above, were too far out and few were brought in that could give the right quality of meat.

For the floating factories the Japanese whaling companies were a more attractive source of business. They had an unsatisfied demand for meat and were ready to bring their own freezing ships and buy meat straight from the factory deck. We negotiated a contract for the delivery of meat from *Venturer* to a Japanese whaling company, but it did not come to fruition. The agreement was subject to the consent of their ministry, and shortly before the deadline the company informed us that the ministry had not given its approval. We found out afterwards that it had made a contract with one of our Norwegian competitors at a lower price. We complained sharply about what we saw as an act of pure bad faith, and set about suing the company for our losses. The Japanese industry, with delicate negotiations in train on the buying of European expeditions, was sensitive about its reputation, and we managed to extract some compensation from the company. Japanese companies made agreements with several Norwegian whaling companies, putting their own gangs on board the factory ships to cut up the whale meat before transfer to the Japanese freezing ship. The Norwegians were much impressed by the skill and organisation of their collaborators.

I went down to the whaling grounds for the fourth and last time in 1958/59. On that occasion the tanker programme allowed me several weeks at South Georgia before visiting *Venturer*. The station was now in a better state than it had been since the war; all the new buildings were completed and numbers were down to 300, a level at which everyone had a full and satisfying job. The manager was now Sverre Akseth, a fine leader, who had come ashore after trying his luck as a gunner. Sadly he died in an accident there two years later, falling into the water between two moored catchers as he was crossing from one to the other – a reminder of the always present dangers of whaling life.

My notebooks from the visit record much detailed probing of the workshops, stores and factory to understand their operations and help with their problems. Often there were complaints and suggestions about equipment and machinery which had been difficult to pass to the office at home during the short summer period between the seasons. On the other side quantities of stores came to light – tools, blankets, paint, ropes, whalecatcher spares, which had been over-ordered in some previous season and needed to be used up or sold. It was satisfactory to be shown on the beach a number of bronze propellers, discarded from the catchers years before because of their singing effect, but worth good money if taken home.

In the factory the new meat extract plant was starting off, though with rather few whales to feed it. The guano shed, which housed the neat stacks of meat meal and bone meal, was now as mechanised as we could get it. Lengths of belt conveyor took the sacks from the plant into the shed. When they were loaded for shipment they passed down similar conveyors to the end of the quays and tipped into barges which took them round to Coronda pier to load by derrick into the holds of the old transport ship *Southern Garden*. It was still a cumbersome business, but better than manhandling the sacks and ferrying them behind a tractor along the main street of the station.

There were still three stations operating on the island that season – Argentina de Pesca in Cumberland Bay, and Husvik and Leith

Harbour in two separate arms of Stromness Bay. Each station was licensed by the Falkland government for seven catchers, an arrangement which satisfied us all. The whale stocks within reach of the island, diminishing there as elsewhere, could not justify a bigger catching effort. It had been temporarily upset in 1957/58 when Tønsbergs Hvalfangeri, the owners of Husvik, decided not to operate. To get some benefit from their catching fleet they chartered some catchers to Pesca, which was then set to operate with a double fleet. When we heard of this plan we reacted strongly, claiming to government that catching licences were specific to stations and could not be transferred from one operator to another. There were good precedents for that view and government was brought to accept it. We thus maintained parity with Pesca. For that season each of our two companies had licences for thirteen catchers, though we did not make use of our full entitlement for the whole season. Husvik came back into operation in 1958/59 and we reverted to the previous régime, but it closed down for good at the end of the following season.

In the last seasons at South Georgia all the stations tried a system of cooperative catching, with gunners and their crews paid bonus not on their own catches but on the average catch of the fleet. The normal arrangement paid bonus to gunners entirely on their own catch, and to crews partly on the expedition's results and partly on their boat's performance. That had always been successful in bringing forward star gunners and getting the maximum catch for the expedition. The fierce competition was never self-devouring and there was plenty of cooperation between the catchers under the control of the leading gunner or the manager. On the South Georgia grounds, with wide searching and long towing, it seemed wise to pool catcher efforts, so that each boat would be available to hunt or tow as needed. The individual gunners had been much closer to each other in performance than in the pelagic expeditions, so the better gunners were happy to come into the scheme on the same level as the rest. The pooling arrangement seemed to be quite successful, though we did not get enough experience to judge whether it got better results than the competitive system.

We were fortunate to have as one of the pillars of Leith Harbour Dr Dafydd Richards, a Welshman from Merioneth, almost alone among Scots, Geordies (Tynesiders) and Norwegians. After a season on *Venturer* Dafydd had found his home at Leith and had just completed a winter when I arrived there in 1958. A sensitive listener and observer, he became the father confessor of many whalers who needed relief and companionship in that close community. He was generally admired, not only as a conscientious and thorough doctor but as a humorous original, bringing some Welsh fantasy to the absorbing business of whaling. When *Venturer* called at South Georgia after his first season Dafydd donned his pinstripe trousers, black jacket and bowler hat, and with rolled umbrella stepped down the gangway to call on his colleague at the hospital. I had his company at a feast of salt herring and boiled potatoes, arranged on one of the tankers by the master, a bible fundamentalist from the Shetlands. Dafydd brought the conversation round to a crux, well known in those circles though new to me. If Adam was the first man and only had two sons, how did Cain find a wife in the land of Nod? We found no solution.

A curious event happened which called on Dafydd's skills outside his normal sphere. As we were eating our tea in the villa one evening the chief steward rushed in and announced that all the pigs were dying. Dr Richards went along to the piggery and found that several pigs were indeed dead, and many others looked very sick. By nightfall there were a hundred carcasses floating out to sea. Richards did some post-mortems and diagnosed arsenic poisoning. The swill was found to have traces of arsenic. Then an empty tin of arsenic rat poison was discovered outside a barrack. Finally it was established that a mess boy who worked with the pig man had drawn a tin of rat poison from the store, and the circle was closed. The boy had taken a grudge against his boss and had revenged himself by poisoning the pigs.

We had some debate on what to do with the culprit. The magistrate at King Edward Cove, who represented government authority on the island, presumably had some judicial powers, but they had not been tested in recent years. In any case the jail had long since been put to other uses. But we certainly had to take the boy away from what remained of the piggery, and he would hardly be accepted in the mess rooms of Leith. Dr Richards came to the conclusion that though mentally unbalanced he was not likely to be a future danger. So the manager shipped him over to Stromness where he worked without incident until the end of the season.

South Georgia was the only area of our whaling which brought us into contact with the Falkland Islands government. As noted above, it licensed our catchers, and it levied tax both on the company and on the income of the whalers. The resident magistrate made sure that the companies obeyed the whaling regulations and conformed with the quite demanding requirements of the Merchant Shipping Acts, otherwise we had virtually no communication with the Falklands. The general view within the company and on the stations was that the Falklands took no interest in the island except to extract taxes, for which they gave no return. Occasionally the governor visited the island, ferried from Port Stanley by a naval ship, though never when I was there. By this time Harold's hostility to governors had faded into the past, and we kept friendly though distant relations with them. One of them kindly sent down to the station an old safe which had been recovered from the remains of the whaling station at New Island in the Falklands which we had abandoned in 1916. I was there when it arrived and there was rare excitement about what it might contain. Keys and oil having failed, a welder was brought along to burn away the lock. When the door finally swung open it was disappointing. There was nothing inside but a Norwegian/English dictionary and a sodden letter book, which appeared to contain copies of letters from the manager to head office. The steward took over the book and put it in a warm place above the baker's ovens to dry. When it was dried, however, the pages proved quite illegible and crumbled to pieces at a touch. I reported this to Harold on my return home. Far from being pleased that I shared his interest in our whaling history he flayed me for having burnt up priceless records.

With whales few and far between there was more time for leisure. This was not altogether appreciated by the whalers, who wanted wall-to-wall overtime to give them a decent pay off. The new cinema was now in full operation, with a tolerable supply of films which were swapped when opportunity offered with those on the floating factories. There was regular football, and the yearly sports meeting with the other stations. We also had a concert party, long-winded but lively. I tried to contribute by putting together a lecture on the history of South Georgia, and gave it in Norwegian and English on successive evenings. It was intended to mark the fiftieth anniversary of Salvesen's arrival at South Georgia. I concluded with the hope that whaling could thrive for fifty years in the future as it had for fifty years in the past. That was, alas, the reassurance of the doctor to the dying patient. But if it made some of the audience more cheerful, including our friends from Husvik, I did not regret having said it.

I sailed out to *Venturer* from Leith on our tanker *Southern Satellite*, delivered new in 1955, and soon to be sent away on long-term charter to the Admiralty. *Satellite* had fetched me from Recife for the trip south. Seeing large quantities of green pineapples on the quays at Recife I got the agent to put a supply on board for Leith Harbour. The ship's refrigerated space was already full of supplies for the station so the pineapples had to be stacked on deck. The heat as the ship sailed down the coast of Brazil largely spoiled them, but enough could be salvaged on arrival to give a good dessert for the station dinner.

On *Venturer* there was as usual plenty of interest, though there seemed very few whales to be found. In the factory the new meat extract plant was settling down. I went out on several of the catchers and visited the rest when they came alongside the factory for bunkering and stores. That season we were working with thirteen catchers, nine full catching boats and four buoy boats. As already explained, the buoy boats were principally intended to serve the catchers, by collecting whales and towing them back to the factory, or holding whales close to the factory when it was moving or there was not space to hitch them at the factory's stern. Some of them had the opportunity to shoot. The senior and best buoy boat carried two half-gunners, who each took half the season as gunner, hunting with the other catchers as much as possible. That was a way of encouraging, testing and training new gunners. If a half-gunner showed promise he might get a full boat the following season, replacing a gunner who had performed poorly. It was easy to bring up good average gunners, but hard to find new stars. The reason was probably not so much a lack of suitable candidates as the increasing importance of the equipment. Large, high-powered boats with good asdic operators counted for relatively more than the skill of the gunner. The corvette catchers, disrespectfully described as 'tin cans' by gunners, would always, with their greater length and sixteen knots, outfish the sturdy but slower Smith's Dock catchers. A gunner who had been a modest performer on the smaller boats would do very much better when he was moved to a corvette. But we only had a few corvettes in 1958/59, five out of eighteen full catchers. Later we bought at low prices some more larger boats which became available from the transfer or lay up of other expeditions. That helped our performance, though some of the purpose-built boats turned out less effective than the converted corvettes. All were still steam-driven.

Even though we had got over our prejudices on the reliability of diesel catchers, it was too late to consider new buildings. When we saw the Norwegian whaling companies building a few we felt that they were either being absurdly optimistic about the future or had special tax reasons for keeping up investment.

Improvement in catching had thus to focus on equipment rather than ships. Asdic was proving an important aid and modifications were improving it. We introduced high-frequency radio telephones, which improved communication between the catchers and factory and could not easily be listened in to by competitors. Course recorders were another refinement. First developed by radio technicians on *Balaena*, the course recorder enabled the gunner to log the bearing and distance between each successive whale he caught in a day so that he could easily retrace his steps to pick them up at night.

During the later seasons we only had one serious catcher casualty. *Harvester* was fishing round the South Shetlands. *Southern Hunter*, passing by the narrow entrance to Deception, decided to go in and look round. As it entered the channel a large Argentinian ship came steaming out, forcing *Hunter* in to the shore. The propeller hit a rock and the ship was disabled and drifted aground. The crew got ashore without problems, salvaging some of the portable gear, but the boat was a total loss.

That was the official account of *Hunter*'s loss, recorded in *Harvester*'s logbook. I found out, nearly forty years later, that it was quite untrue. There had been an Argentinian ship coming out behind *Hunter*, but it was not involved in the grounding, which was due simply to careless navigation. The true story came to me from one of our gunners, later a navigation teacher at the Tønsberg maritime school, and was confirmed by a Scot who as a messboy had been in the crow's nest when the ship struck the rocks.

Our helicopter units were still working in the 1958/59 season, but our earlier enthusiasm for them had been dampened. In 1956/57 we had lost an aircraft on the whaling grounds through engine failure when taking off, and the next year one was damaged in England when flying to join *Venturer* before sailing. During that season the remaining *Venturer* aircraft had to land on the sea due to fog and could not be recovered. Fortunately there was no more loss of life. For 1958/59 we fitted floats on the aircraft to make ditching on the sea less dangerous; we also left one of the expeditions with only one helicopter. Thereafter there was a gradual run-down, as the risks seemed too high and the benefits uncertain. By 1960/61 we had one helicopter on *Harvester*, nothing on *Venturer*. The following season we carried no aircraft and disbanded the unit.

It might seem odd that among the many technical improvements in the catching of whales there was no progress in the basic technology of harpooning and killing. The harpoon gun, heavy harpoon and rope system still in essence followed the techniques developed by Svend Foyn eighty years before. Many refinements had been introduced over the years in the design of the equipment. The guns became breech loaders instead of muzzle loaders (though our leading gunner Hans Abrahamsen insisted on a muzzle loader for himself right to the end). There were improvements in the harpoon design, the traditional pointed grenade being superseded by one with a flat saucer-shaped end which was not deflected when it hit the water. The development of the nylon foregoer, combining great strength with elasticity, reduced breakages in the harpoon lines. The increased size and power of the catchers made it easier to bring in a wounded whale quickly and despatch it with a second harpoon. Nevertheless the method of taking whales remained clumsy and sometimes inhumane. A wounded whale might suffer for half an hour or longer before it was finally killed. Ropes could still part or harpoons break their hold, freeing the wounded whale for a slow and painful death, or leaving a killed whale to sink and be lost. As early as the 1920s experiments were started with electric harpooning which would kill the whale on impact with a single electric shock. These were carried on in the 1930s by Norwegian companies with some, but no convincing, success. After 1945 there was renewed interest in developing an electric system. This time our British competitors, Hector Whaling, were involved, and their chairman, Sir Vyvyan Board, went to try it out himself on the whaling grounds.

The industry's search for a more effective way of killing whales was strengthened by animal welfare bodies who campaigned for more humane killing. One of their leaders was Dr Harry Lillie, who had been doctor on *Harvester* in 1946/47. The matter was also on the IWC's agenda and scrutinised from time to time. But the technical problems of electric whaling proved too difficult to overcome without seriously damaging the effectiveness of hunting, which was the main concern of the whaling companies and their gunners. It was found that the electric shock did not always kill the whale, which was sometimes only stunned and came back to life. A shock sufficient to guarantee death would have had to be much stronger, with attendant risks to the gunner and catcher crew. There was also difficulty in the transmission of the current, since it had to go by a cable which was part of the foregoer. This could be made to stretch with the foregoer but its incorporation weakened it, and the link between the cable and the harpoon was not secure enough. The champions of electrical harpooning claimed that it would have succeeded but for the shortsightedness of the companies and the conservatism of the gunners, but the companies, whose interest was naturally in better and quicker killing of whales, would certainly have gone further with electric whaling if they had seen enough promise in that direction. They were not, however, prepared to promote a less effective method on humane grounds alone. The causes of failure of electric whaling lay not so much in deficiency of will, or even of humane feeling, on the part of the whaling companies, but rather in the immense difficulty of applying complicated technology to the violent and unpredictable setting of the whale hunt.

We in Salvesen had carried out some trials at Leith Harbour in the 1930s. After 1945 we did no more than follow Hector's trail with interest. Harold liked being a pioneer, but sensibly preached against trying to be a pioneer in everything. In an open industry like whaling, where everyone would be able to take the benefits from innovation, it was sensible that the risks of innovation should be spread. We pioneered in asdic and helicopters. Others could take on electric whaling.

Concern for the humane side of catching whales came out in a debate in the House of Lords in 1958 led by the Bishop of Portsmouth. The bishop had himself served in the Antarctic and argued for more vigorous steps to develop electric whaling. The honour of the whalers was ably defended by Lord Geddes, a director of P & O and, for a short period after the war, a partner in Salvesen. The debate, as always in that chamber, was pursued in measured terms without emotion, and criticism of the whalers fell well short of the vehemence of present-day animal rights campaigners. Occasionally the press took up accusations of cruelty, making little public impact but bothering the whalers, who were proud of their profession and had never seen themselves as being party to the deliberate infliction of pain.

Apart from that the whaling industry did not raise much interest in Britain. The romance of Moby Dick did not extend to the industrial era, and it was thought of as a dirty and rather shameful occupation, pursued bleakly beyond the margins of the civilised world. It suited us at the time to discourage any enthusiasm for whaling stories from newspapers; we had nothing to hide, but we knew from experience that any journalist would get the facts wrong and be unfriendly to the industry. Harold, for whom accuracy and truth rated as high moral principles, got furious about press articles and fired off angry letters to editors listing the mistakes. His complaints were of course ignored.

In Norway, by contrast, whaling was a respected industry, close knit with the local communities. The fortunes of the whaling expeditions were followed with as much interest as the local football teams, and international whaling negotiations rated editorials in the national press. We cultivated publicity there as eagerly as the Norwegian companies.

The difference in attitude between Britain and Norway is well illustrated by the behaviour of the British and Norwegian governments in the IWC after we came out of whaling. The British government turned overnight from conscientious support of its industry to an ultra-conservationist stance. It joined the Americans in pressing for a complete ban on whaling, regardless of the state of the whale stocks. A veil was drawn over the past, which it preferred to forget. The Norwegian government continued to fight for the industry even after its Antarctic interest had gone, and championed the cause of its small whale fishermen against considerable international pressure and strong opposition by 'green' groups in Norway.

Publicity occasionally gave us some diversion on the whaling grounds. At the end of 1956 Prince Philip visited the Antarctic on *Britannia*. He called on *Harvester* and was lunched on pea soup and salt beef. A few days later *Britannia* berthed at Leith Harbour for a tour of the station. On the way north the ship passed some of the Leith catchers at their work and the royal party was able to see the chasing and shooting of several whales.

In an earlier season we relaxed our suspicion of public exposure by allowing a film crew to spend some time on *Harvester*. Background film was required for the screen version of Hammond Innes's novel *The White South*, which came out as *Hell Below Zero* (the original title being judged unsuitable for American audiences). Innes's novel was an excellent thriller based on Antarctic whaling, with his unique mixture of heroism, hardship and romance. His evocation of whaling is almost flawless, and his characters, except for the girl gunner, were uncomfortably close to some of the leading whaling personalities. He had put together much of his knowledge of whaling at a station in north Norway. But he went wrong in supposing that the gunners of catchers visited the factory ship by rowing boats, which they never did. The catcher berthed alongside the factory, and crew members came and went via the traditional coal basket hoisted by a derrick.

The red, white and blue twin funnels of *Harvester* came out well in the film. The ship's whalers provided stand-ins for Alan Ladd, and even for Jill Baker, seen behind the gun of her catcher. Rumours that the stars would appear in person provided some excitement before they were scotched.

I came back at the end of the 1958/59 season via Cape Town, being dropped there by *Southern Satellite* on her way north. We had had little to do with South Africa since the lay-up of our station at Saldanha Bay, west of Cape Town, in 1931, but it was an important centre for whaling. The South African company Union Whaling had operated *Abraham Larsen* from there and had a very successful land station in Durban, while Hector Whaling had taken up whaling from Saldanha Bay. Cape Town was also used for overhauling whalecatchers. As our centre of operations was South Georgia, Cape Town was too far east to fit in with them.

I took the opportunity to call on the captain of the new *Willem Barendsz*, just back from the whaling grounds. I envied him his magnificently large and well-appointed ship, though noting that it was sadly deficient in the by-product plants with which we would have crammed the space. The Norwegian government had continued to ban their gunners from taking employment with the Dutch, but that did not stop the Dutch from recruiting them. It meant giving up their domicile in Norway and moving to South Africa, where the attractions of higher pay and lower tax were considerable. Among them were some of our ex-gunners, whom I was pleased to meet again on *Barendsz*. One of them I would have preferred to avoid, as he had bouts of paranoia which made him a dangerous companion after a drink. After I had seen his vicious handshake in the captain's cabin, when he nearly broke the hand of an innocent customs man, I steered clear of him before worse happened. A round of Cape Town night spots with him would have been certain disaster.

Six: 1959-1963

We were now entering the end game of Antarctic whaling. The acceptance by the USSR in 1958 of 20% of any total quota fixed by the IWC had opened the way to a general agreement on national quotas, but the resignation in 1959 of Norway and Holland from the Convention had effectively removed the catch ceiling, while whale stocks were being depleted at an alarming rate. The best rough measure of that was in the catch per day per whalecatcher. In 1958/59 it was 0.94 blue whale units; next season it was 22% less, at 0.73, and in 1961/62 it went down to 0.51, a drop of 46% in three seasons. If correction was made to recognise the increasing efficiency of the fleet the figures would have been even more striking. These statistics were shaky ones because the USSR figures could not be trusted, but there could be no doubt that whales were fast running out.

After the signature of the 1958 agreement, with the USSR accepting a 20% share, the next step was to allocate the remaining 80% between the other four whaling countries. The Norwegians and British hoped for a quick settlement as a prelude to reducing the overall catch ceiling and protecting the remaining whale stocks. But the other two countries, Holland and Japan, did not feel the same sense of urgency. The Dutch still did not believe that whales were being overfished, and maintained stoutly that for historical and national reasons they deserved a higher quota than anyone else. The Japanese were not in a hurry to accept a quota. They were still doing well under the free competition system, and they could see the Europeans selling to them or going out of business in the coming years.

The Norwegians led from a weak hand. They claimed a quota based on past catch results from all nine of their expeditions. Japan and Britain thought that too generous. Three of the Norwegian factory ships were small and old, and on the verge of being laid up or scrapped. If they were going to be withdrawn anyway there was no justification in giving them a quota which would then be passed to the larger expeditions. The freezing of shares of a competitive market, as brought about by a quota system, only works if the parties are satisfied that it will give, with less bloodshed, the results that would have been reached by free competition. The proposals from Norway did not seem to reflect this.

The Dutch intransigence appeared to rest on their misguided confidence that whales were still abundant. They also argued that unless *Barendsz* got a quota far higher than expeditions of similar capacity Holland would soon be squeezed out. If the overall quota was brought down to 10,000 units or lower, the quota for each expedition would be too low to allow it to operate economically. Each country would have to allocate its units among fewer expeditions, getting the companies to arrange lay-ups and mergers. The Norwegians and British were quite clear that this would have to happen. The Dutch with their single expedition had no such way of ensuring the survival of their whaling at a lower level. Recognising this, and believing that the Dutch were the main stumbling block to a national quota agreement, the Norwegian companies negotiated in 1959 to buy *Barendsz* and take the Dutch out of whaling. The British and Japanese companies were ready to participate. The Dutch government, which had given full and expensive support to its whalers over the years, now saw in the proposals an honourable exit, and promoted the discussions. They looked like succeeding, but eventually the Dutch company turned them down.

If the Dutch had been out of the way, a national quota agreement would probably have emerged in 1959. The Norwegians would have had to accept a quota which reflected the capacity of their six viable expeditions with some, but not full, extra provision for the three old expeditions that were already close to lay-up. The Japanese, who wanted more for each of their expeditions on the grounds that they were more efficient in catching than the Europeans, would have accepted figures close to the British and larger Norwegian expeditions. They fully intended anyway to increase their quota by buying up the Europeans. The British would have been content with a quota that was not less than the others. Harold tried to strengthen the British position by claiming that expeditions with high by-product production should get more, thus boosting the British, and even more the Japanese, share at the expense of the Norwegians. But it would have been very difficult to introduce by-products as a new criterion in addition to that of basic catching and processing capacity. By-product assessment would inevitably have brought in comparative economic performance, which no one would have thought a proper basis for quota allocation, even though in the long run under free competition it would be the determinant for relative national positions. In any case the by-product scene was moving fast; most of the Norwegians were looking at meat extract plants or the potential of selling meat to Japanese ships. Harold realised all this, but he was concerned to brace up the British negotiators in arguing our case, to '*oppstramme*' them, as he expressed it in Norwegian. We had the Fisheries Secretary with capable supporters to represent us, but Norway fielded a cabinet minister, Gustav Sjaastad, a man of great experience and ability, committed to defending his country's still dominant position in whaling. The principle of full utilisation of the whale had been an enlightened aspect of UK government policy almost from the beginning of Antarctic whaling, and the British companies, Hector as well as Salvesen, had supported it with enthusiasm. In a negotiation which concerned a world resource, with implications far beyond the immediate interests of the whaling countries, it was appropriate to lodge some moral claims as well as practical ones.

So no national quota was reached for the 1959/60 season. As noted above, Norway and Holland resigned from the Convention, and each country decided its own catch figures for 1959/60. Nor was any progress made for the following season. The Norwegians still claimed a quota to cover fully their older expeditions. The British claimed parity with the modern expeditions elsewhere. The Dutch held out for special treatment and an unrealistic share. The Japanese and Russians were evasive. All that emerged was a loose agreement that the unilateral quotas which countries had adopted for 1959/60 should not be exceeded for 1960/61. In summer 1960 *Balaena* was sold to the Japanese, leaving *Venturer* and *Harvester* as the only two British expeditions. In the spirit of the quota negotiations *Balaena* took with her to Japan part of the British adopted quota, though not a full third share.

A substantial difficulty in completing a quota agreement was the absence of Norway and Holland from the Commission. Norway had resigned from the Convention because she found the free competition system intolerable and hoped thereby to press the other whaling countries into coming to a quota agreement, but her action had unfortunate consequences. The Russians indicated that they could not confirm their part of a national quota agreement until all the countries concerned were back as members of the Commission. This also applied to the International Observer Agreement, which the other whaling countries hoped could be negotiated at the same time. The British were also keen to get Norway back. The Norwegian resignation, by effectively raising the total catch ceiling, had released the British companies from the trap of free competition within a low catch ceiling, but they were afraid that the Norwegians might be tempted to drop other restrictions of the Commission, particularly the season limitations, as the Dutch had already done. The non-whaling members of the Commission, led by the USA, were dismayed at the departure of what was still the leading whaling nation. If Norway did not come in again soon what future was there for the IWC and the international system of regulation built up since its inception in 1946?

The British delegation to IWC, led by the Fisheries Secretary, Basil Engholm, took the lead in getting the Norwegians back. He persuaded the Commission at its 1960 meeting to suspend the 15,000 unit catch ceiling, which was still in force, for the following two seasons, so that the Norwegians could come back without fear of being trapped within the 15,000 units. The British government also proposed at the same IWC meeting the setting up of a small expert committee to examine and report on the whale stocks and stressed how important it was that all the whaling countries should support it.

There was fierce debate in Norway on whether Norway should go in again. The industry thought that the only way to force a quota agreement was for Norway to stay out until a satisfactory settlement was reached. The government was more worried that the Russians, and the Japanese too, would cease to play in the quota negotiations if Norway stayed out. So Norway did rejoin at the end of 1960, but a year later she gave notice of withdrawal again. National quotas had still not been agreed, and she was afraid of being trapped once more in free competition under a low catch ceiling after the two years' suspension of the limit had run out in 1962/63.

Holland had been subject to similar pressures to come back in, but took no notice, happy to pursue her own course to the end.

It seems to me, looking back on those complicated and wearisome negotiations with thirty years' hindsight, that Norway's departure from the Convention, subsequent re-entry, and further notice of resignation had little effect on the progress of the quota negotiations. Someone had to resign in the first place to blow up the failed free competition system. After that it was probably a matter of indifference to the other whaling countries whether Norway was in or out, provided she kept to the regulations. The Russians might well have gone back on their 20% agreement at any time through some unpredictable change of policy. They remained content with that figure, no doubt reckoning that any agreement limiting catch which others would adhere to and they could break was worth encouraging. They were certainly not going to put into effect any International Observer Agreement, as was subsequently shown. The Japanese and the Dutch were not affected by Norway's position. The Dutch resolutely stayed outside anyway, while the Japanese rated their growing superiority a much more important negotiating factor than the formal question of IWC membership.

It was the swiftly falling catch, shown in the 1959/60 and 1960/61 seasons, that brought the whaling countries back to serious national quota negotiations at the beginning of 1961. They now came quite quickly to a provisional agreement which gave, in percentage terms of the overall quota, 33% to Japan, 32% to Norway, 9% to UK and 6% to Holland, the remaining 20% going to the USSR in accordance with the 1958 agreement. That was for all except Holland a reasonable distribution, reflecting capacity and past performance. Harold had hoped for 10% for the UK and was furious when the British negotiators accepted 9%, but that was in line with the Norwegian figure. Two of the three smaller Norwegian factories were still operating, so the argument no longer held that the Norwegian fleet was getting extra units for ships that would be laid up.

The bargaining hands of the Norwegian and British governments were weakened by the knowledge that the troops they led were not pledged to undying loyalty. *Balaena* had been sold to Japan and Salvesen had very nearly laid up *Venturer* for 1960/61. The Japanese were knocking on Norwegian doors, where there was no reluctance of owners to sell. So the governments, though they of course hoped to give their companies a stable future by a satisfactory quota agreement, justifiably felt that they were simply negotiating national quota rights to give owners a windfall profit when they sold their ships.

The Japanese showed themselves remarkably accommodating. Although their performance since 1958 had been substantially better than the other operators, that was not given the weight which they might have claimed. The figures agreed for the three countries turned out extremely close to those originally proposed by Sjaastad in the 1958/59 negotiations.

The stumbling block was Holland. The Dutch had moved sufficiently to recognise that they would never be able to catch the exorbitant number of units that they had claimed in 1958, but they still insisted on special treatment and, encouraged by their scientists against the growing weight of evidence, they still claimed that the fin whale stocks were in reasonable shape and there was no need to reduce catch. The final proposal allotted the Dutch 6% for their single expedition, compared to 4.5% for each of the British expeditions. This they turned down, claiming a further ninety units as the minimum they could accept. Negotiations on whether or how they could be accommodated went on during the summer. A final formula, which virtually accepted their claim, was arrived at by October 1961, but the agreement of the USSR was required. The Russians kept to their formal position that they would not sign till Holland had rejoined the Convention, while the Dutch refused to rejoin until the agreement had been signed. That took more time to sort out. So a further whaling season passed with quotas declared unilaterally and no catch ceiling. Eventually, in June 1962, the agreement was signed by all five parties and Holland came back to the Convention.

So the Antarctic national quota agreement, which we had intended to negotiate for the 1959/60 season, following the acceptance by the USSR of 20% in autumn 1958, was not completed and put into effect until the 1962/63 season, three years later. During that period, with the catch ceiling removed, irreparable damage had been done to the whale stocks. Instead of being brought down to 10,000 or 11,000 blue whale units, as the IWC Scientific Committee had recommended, the total catch in those three seasons was an average of 15,700 units, well over the 14,500 catch ceiling applying five years earlier.

There was no lack of voices inside and outside the Commission to point out the disastrous consequences of this reversion to anarchy. Dr Remington Kellogg, veteran leader of the USA delegation, who had made the conservation of Antarctic whales his life's work, declared his intense distress at the increasing ineffectiveness of the Commission. He was supported strongly by Canada and New

Zealand, sufficiently detached from Antarctic whaling to speak objectively. The whaling countries listened to these voices, but the strongest pleas could not help them solve their immediate negotiating problems.

The delays which strung out the negotiations for three years can be analysed and understood, but it is difficult to believe that they would have been so protracted if all the parties had felt the same urgency to agree as the Norwegians and ourselves. The Dutch, as has been seen, recognised no such urgency and pushed their obstinacy beyond the limit. The Japanese probably were less sensitive to the decline of the whale stocks and could see their position strengthening as the negotiations dragged on. Over the whole scene loomed the shadow of the USSR, which had enabled negotiations for national quotas to be set in train by accepting 20%, but would go on cheating. The new Russian fleets, catching freely, could make a nonsense of stock conservation policy. That prospect did not encourage the other whaling countries to move decisively.

There would have been more confidence in the future of whaling if an international observer scheme had been adopted, posting observers from other countries on whaling factories to ensure that the rules of the Commission were kept. This had been a matter of concern for the Commission since 1955. Onassis had been able to fish with complete disregard for the IWC although *Olympic Challenger* sailed under the flag of Panama, a member of the Commission. The Russian infringements, sometimes disguised as 'taking whales for scientific purposes', were being increasingly exposed. The USSR had also maintained its habit of overdeclaring *Slava*'s catch, no doubt to enhance its position in any quota negotiations. The figures published gave a catch 50% or so above the best European and Japanese expedition, far above *Slava*'s processing capacity.

An international observer scheme was put forward at the 1957 meeting of IWC and commended to the whaling countries for a separate agreement, as the IWC had no powers to introduce one itself, but the Russians were not prepared to negotiate. It had been intended that the international quota negotiations of 1958 should be linked to such a scheme; indeed part of the reason for Norway and Holland leaving the Convention was to force through an observer agreement. But with the Russians still showing no interest that condition was dropped. When the national quota agreement came into effect in June 1962 the observer agreement was still in limbo.

While free competition with an overall catch ceiling was operating the only communal way that companies could improve their economics was by agreement between themselves to limit the number of whalecatchers per expedition. If applied universally that would leave the competitive position unchanged. The catch ceiling would be reached later, and each expedition might be expected to take the same number of units with fewer catchers and lower costs. Although the Russians would not cooperate the other four countries found it worth while to negotiate catcher limitation agreements, and did so in 1953/54 and 1954/55. There was a break in 1955/56, but the following three seasons saw new agreements which reduced significantly the number of catchers per factory. In 1955/56 the average had been 13.5 per factory; the next season, with a limitation agreement, it was reduced to 11.3.

In the seasons after 1958/59, with national quotas agreed or declared, there was no point in having general catcher limitation agreements. Maximum catch quota was fixed for each expedition, and the catching power deployed by its competitors could have no effect on either its own catch or the length of the season. It might have been expected that the individual companies would increase their catcher fleets to give a better chance of catching the high quotas which had been declared for the 1959/60 season. That did not happen. Companies could now operate the full permitted season, which would not be cut short by the overall catch ceiling being reached, so with a lower number of catchers they had better prospects than previously. Moreover as whales got scarcer the additional contribution of extra catchers was less. Falling whale oil prices reduced the value of each extra whale, though that was balanced by the growing value of by-products. On the other side, the cost of an extra whalecatcher was not only the direct running costs of the boat. If the fuel oil consumption of the expedition could be reduced by having a smaller catcher fleet it might be possible to reduce the tanker service needed to supply it, with considerable savings. The two Salvesen factories operated with twelve and thirteen catchers in 1958/59. In 1962/63 *Harvester* for her last season fielded only nine.

The catch of our two floating factories and of Leith Harbour in 1958/59, the last season of free competition within the overall quota, was poor, the lowest recorded since the war. As usual, unusually bad weather was blamed which did seem to have exaggerated the effects of declining stocks. The adoption of national quotas in 1959/60 encouraged us to continue at full strength, and the results in that season were somewhat better. Leith Harbour continued at a very low level, but lack of catch was compensated by high meat extract production. Overall financial results however showed little better than break even. There were now strong economic reasons for laying up one of the floating factories: word was sent out to *Venturer* on her homeward voyage that we did not expect to operate the expedition in the 1960/61 season. It was a depressing message for the crews, though they had been prepared for it three years previously when *Venturer* was nearly bought by a Japanese company. But the lay-up did not come to pass. In July 1960 our British friends and competitors, Hector Whaling, sold *Balaena* to Japan, which changed the whaling picture for us in several ways. We could strengthen our position by bringing in good people and more powerful catchers from *Balaena*; in particular Per Virik, *Balaena*'s outstanding manager, became available to us. There was also a more general consideration that the loss of two expeditions at the same time from Tønsberg, where most of our own and Hector's Norwegians came from, would be a heavy blow, and that if we were now the sole standard bearers of British whaling we had some duty to keep it going as long as possible. The only disappointment with the *Balaena* sale was that we failed to get hold of Finn Ellefsen, *Balaena*'s top gunner, and the best in the Norwegian and British fleets. Ellefsen had started whaling as an AB with our own star, Hans Abrahamsen, and might have been disposed to follow Virik into Salvesen, but one of the Norwegian companies offered him a brand new powerful catcher. We could not compete with that.

So the three expeditions went out again in 1960/61. Catch results turned out very good, particularly for *Venturer* under Virik, and financial results of the whaling operations came back into the black.

In the spring of 1961 we had a new approach from Japan to buy one of the expeditions. There was a lively interest from two competing Japanese groups. Bill Hollyer, our partner in charge of sales, had been contacted through our whale-oil selling brokers, David Geddes, and had a tricky job in keeping them both at arm's length. There could be no doubt that it was right to sell. The price at £2.7m. was still a high one for a factory ship which might soon have no whales to catch. It was of course not the ship itself that was of value but the quota rights, which under the quota agreement still being negotiated would transfer from the UK to Japan. The expedition chosen for sale was *Venturer*. The two expeditions had always been equal in strength and performance, but *Venturer* was a year older than *Harvester* and had slightly smaller engines, so she had to go first. Other Japanese interests were also negotiating at the same time to buy *Kosmos III*, one of the large Norwegian postwar factories. The Japanese government however feared that taking over two more expeditions together might provoke resentment from the other whaling countries and prejudice Japan in the quota negotiations, so it gave permission for only one expedition to be transferred. *Kosmos III* went to Japan and the *Venturer* sale was dropped for the time being. For the third time *Venturer* was reprieved.

We had expected to wind up whaling at Leith Harbour at the end of 1960/61. Catch had been rather better than in the previous season, mainly because Husvik had closed down and only ourselves and Pesca were left fishing from the island. But despite high meat extract production the station was losing money and there was no prospect of any improvement. When the *Venturer* sale was agreed we prepared to give Leith another season to avoid too drastic a run down of our operations in a single year, but with the cancelling of the *Venturer* sale we could proceed with the original plan. So Leith Harbour whaling closed down in 1961, having operated continuously since 1909, except for 1932/33 and the three last seasons of the Second World War. It did later come back to life for two seasons under a Japanese whaling company (see p. 164).

With the end of whaling operations at Leith it was no longer sensible to retain the station as a winter repair base for the catcher fleets of the floating factory. We had for some time been worrying about the high costs of repairing the fleets at Leith and Stromness with expensive and increasingly reluctant overwintering crews, instead of bringing them home to the UK or Norway between the seasons. The closure of whaling from the island meant the end of the Antarctic repair base. We managed to find a suitable yard at Melsomvik outside Tønsberg, which had been used by one of the older Norwegian expeditions, *Pelagos*, until she was scrapped. We shipped home the considerable stocks of spare parts and materials held for the fleet, together with some of the machines. The two floating docks at Stromness and the various workshops, including the fine boiler and plating shop at Stromness, were abandoned. For the remaining two seasons the stations were left with a handful of men, living on a catcher moored at Coronda pier, to help the floating factories when they called at the beginning and end of the season and to do any basic maintenance they could.

The 1961/62 season gave catch results for our expeditions well below the previous season. There was also a crisis in the whale oil market, where prices, which had almost never gone below £70 since the war, dipped to as low as £40. The break arose from the appearance of large and, it seemed, increasing quantities of fish oil from Peru. (My contacts at that time with the Peruvian operators turned our minds to the fishmeal industry and led to a new chapter in our company history.) So our expeditions made an operating loss. Fortunately the whale oil price bounced back again at the end of 1962, partly because of the fright everyone got during the Cuban missile crisis when the USA and the USSR came close to war.

After the failure of the *Venturer* sale negotiations in 1961 there was a new approach by a consortium of Japanese companies. The price was now £2.15m. This time the contract was approved by the Japanese government, and the ship was handed over in June 1962. The Japanese had no intention of operating *Venturer* themselves, but only of getting part of the UK quota. Britain under the final quota agreement had been allotted 9% of the total quota for our two expeditions and we transferred 4% with *Venturer* leaving 5% for *Harvester*. The ship herself sailed to Japan to lay up and was scrapped a few years later.

In summer 1962 the national quota agreement at last came into force. The final figures agreed were 20% for the USSR, 41% for Japan, 28% for Norway, 6% for the Netherlands, and 5% for the UK. Those figures allowed for the transfer of *Balaena*, *Southern Venturer* and *Kosmos III* to Japan, each taking with it a share of its national quota. In fact Japan settled for rather less than she might have claimed, partly because after the purchase of *Balaena* and *Kosmos III* one of the older Japanese factories was withdrawn from the fleet.

The overall quota for 1962/63 was restored at 15,000, as it had been in 1958/59, ending the three seasons of unilaterally declared national quotas which in the past two seasons had added up to 17,780 units. But the position of the whale stocks had worsened in the intervening years. In 1961/62 the Antarctic fleets, despite the generous catch ceilings, had barely succeeded in taking 15,000 units. One of the triumphs of the IWC, in a period of otherwise total frustration, was that it had agreed in 1960 to appoint an expert committee of three to report on the whale stocks, using the best techniques and information available, with the intention that the Antarctic catch should be brought into line with its recommendation not later than the summer of 1964. The expert committee, the Three Wise Men, did not report until after the 1962/63 season. Meanwhile the Norwegians and British fully realised that the current catch was far too high. That view was not shared by the Dutch, who still clung to their optimistic assessments, and only reluctantly supported by the Japanese, who were still catching the full quotas they set themselves, helped by their continuous investment in new and more powerful whalecatchers.

The now acute worries about catch were reflected in the line-up of expeditions for 1962/63, totalling 17 factory ships and 201 catchers, against 21 and 261 in the previous season. While the Japanese maintained seven expeditions and the Russians four, Norway reduced from seven to four. *Southern Harvester* represented the UK and *Willem Barendsz* the Netherlands. Results followed the now established trend downwards and none of the European expeditions came anywhere near to reaching their quotas. The index of blue whale units per catcher day's work, in spite of a smaller and more powerful catcher fleet, continued to go down. *Harvester*, strengthened by concentrating the best gunners and catchers in a single expedition, did best of the Europeans, but her production of 80,000 barrels was still modest.

The final adoption of the national quota agreement had left still uncompleted the International Observer Agreement. Attempts had repeatedly been made during the quota negotiations to make the one dependent on the other, but with no response from the USSR that had to be dropped. In early 1963 the Russians surprisingly took it up and invited the whaling countries to a conference in Moscow in May with the aim of signing an agreement. The UK delegation consisted of Ronnie Wall and Ian Graham from the Ministry of Agriculture and Fisheries, with myself as industrial adviser. Wall was a hardened IWC man, who had been chairman of the Commission in 1959 while he was Fisheries Secretary. His association with the IWC continued well after the UK had left whaling. Ian Graham had no less experience of the workings of the IWC and other fishery bodies. In his years of international negotiation he had formulated what he called Graham's Law, to the effect that no scheme for the protection of an endangered species ever became effective until the last surviving member of that species had been eliminated.

The three of us went to Moscow in some euphoria. At last it seemed the Russians had been converted to honesty and conservation. The first two days were May Day holidays, and we had to wait patiently while endless processions of tanks and guns rumbled through the Moscow streets. On the Wesnesday morning we started work. By Thursday evening a satisfactory draft agreement had been hammered out and we all went to the opera, *Prins Igor*, at the Kremlin theatre. But we were not to get away with it so easily. When we assembled on Friday morning the Russians announced that they now could not accept the draft agreement and negotiations would have to start again. We worked intensely through Friday and Saturday, finally reaching agreement on Sunday ten minutes before we had to leave for the airport to catch the last flight to London. Ian Graham was a supreme drafter, skilfully converting fragile understandings into durable clauses that were not entirely ambiguous. His wordings, little different from the rejected draft, provided the final accepted version. We eventually realised, as we should have done earlier, that the whole negotiation had been stage-managed by the Russians, the sudden switch from friendly cooperation to aggressive suspicion being calculated to confuse and weaken the other negotiating parties. The triumph of returning home bearing, like Neville Chamberlain, peace with honour, was equally shortlived. It soon became clear that although the USSR had signed an International Observer Agreement it had no intention of implementing it. It successfully delayed action, on technical and procedural grounds, for a further nine years. By the time the agreement came into force Antarctic whaling was virtually at an end. Graham's Law was vindicated.

My main role in Moscow was to be business manager of our delegation. One job was to try to get breakfast served at the hotel in less than two hours so that we could appear at the negotiating table with something in our stomachs. If we had been less trusting we would have brought our own canned sardines, as frequent travellers to Moscow often did. Another assignment was to find a restaurant where we could entertain the Norwegian delegation to dinner in reasonable privacy and with decent food. The British embassy people directed me to a Georgian restaurant, the only suitable place in Moscow, they said. I called on the proprietor, who clearly operated his business outside the official system. He was most obliging in planning the menu and wines with me, and the party was a great success. Payment was of course in hard currency. Roubles were of little interest to him, still less the tourist vouchers with which foreign visitors tried to get access to his privileged tables.

There was also a dinner given by the Soviet Fisheries Minister Ishkov for the British delegation. During that dinner I got into trouble with Ishkov for failing to drink my glass of vodka in the initial round of toasts. I was recovering from some prostate trouble and was off alcohol. Ishkov spotted that I had not drunk and let forth a volley of abuse. Brushing aside my feeble excuses he roundly declared that he would arrange to get me to Leningrad where the sailors of his fleets would throw me into the sea. This sounds rather improbable, and I may have misheard the interpreter translating from the other end of a long table, but it seemed to be the gist. My companions rightly kept quiet, and the incident vanished without trace. My vodka glass remained untouched. Ishkov, a burly party tough, commanded Soviet fisheries for a long time, over twenty years, and was presumably responsible not only for the USSR's successful sabotage of the IWC, but also for its disastrous policies in deep sea fishing.

In the early summer of 1963, shortly before the annual meeting of the IWC, the final report of the Committee of Three scientists was published. Its work on the Antarctic whale stocks was far more thorough than anything that had been done before. The catching rates of the most recent seasons provided much of its material, and it applied to them new statistical techniques backed by computer programmes. The three scientists, D. G. Chapman, S. J. Holt and K. R. Allan were all experienced in the dynamics of fish populations, and they were ably supported by J. A. Gulland, the population expert of FAO. Their report had the ring of authority. Whereas previously whaling countries and other scientists could cast doubt on the stock estimates produced by the IWC's scientific committee, the Committee of Three's report could not be credibly challenged.

The report found that as at the end of the 1961/62 season the main blue whale stocks were at a very low level. Less than 200 a year could be taken without further depletion. Fin whales had a sustainable catch of less than 7,000 a year. Those figures would be further reduced if the catch of 1962/63 was taken into account. The actual catch of blues in that season was well over 200. (There was an additional catch of pygmy blue whales, considered to be a separate stock.) Actual catch of fin whales was 18,668. The figures recommended for sustainable yield by the committee were thus a long way below those levels. To allow stocks to rebuild the catch limits would have to go even lower.

Blue and fin whales were not the only catch in the Antarctic. There were small stocks of humpbacks, also depleted, sperm whales, which could be taken outside the baleen whale season, and sei whales. Sei, which were similar to fin whales but smaller, came increasingly into the catch as the bigger whales became scarcer. The committee had not studied them, but it observed that they were probably already being taken at over the sustainable yield.

The right policy to save the whale stocks would have been to stop the catch of blue whales (except for pygmies) immediately and bring down the fin whale catch to not more than 5,000, taking account of the effects of the 1962/63 catch.

The Commission did go some way to carrying out the urgent advice of the Three Wise Men. Blue whale catching was stopped, except for the pygmies. So was the taking of humpbacks, which the committee found had been reduced to very low levels. When it came to the main stock for fishing, the fin whale, there were strongly divided views among the whaling countries. The line up followed the split that had opened up over the previous five years of negotiation on quotas, with the Russians, Japanese, and Dutch determined to keep their whaling industries going whatever the consequences. After the Committee of Three report they could not oppose a drastic reduction in catch ceilings, but they were not going to accept anything near the recommended cuts in catch, which would mean laying up two thirds of the Russian and Japanese fleets and putting the Dutch out of business. The Russians faced political dangers. How could the Fisheries Minister, who had promoted a vast expansion in the Russian fleet, go back to his government and report that there were no whales for these expensive new expeditions to catch? The Japanese were still finding no difficulty in catching their share of the present 15,000-unit quota, so were reluctant to accept completely the findings of the Three Wise Men. The Dutch, no longer able to find shelter in the discredited views of their scientists, were determined to keep their national banner fluttering to the end.

On the other side were the Norwegians and British. Having led

ABOVE A sperm whale at Leith Harbour, South Georgia, c.1914. BELOW Floating factory *Neko* (1911–24) and transports at Leith Harbour, 1917.

Whalecatcher *Sonja* (1910–24), probably in Leith Docks, Scotland, c. 1910.

f.f. *Salvestria* (1929–40)

Whales seen from a whalecatcher

Working up whales on the plan of a floating factory

f.f. *Southern Venturer* (1945–62). Painting by George McVey, 1956.

Whalecatcher on voyage home from the Antarctic after the season

Whalecatcher chasing off South Georgia, by Frank H. Mason, 1952

the initiatives for conservation for so many years they could hardly do other than support a total catch level brought down to the figures set by the Three Wise Men. The non-Antarctic whaling countries, led by the USA, were of course strong supporters of the same line. They had watched with growing dismay the dwindling of the whale stocks and saw this as the last opportunity to save them. But whatever the voting strength of that group the decision on catch limits had to be made by the whaling countries, and in consensus. If a low catch ceiling was set by a majority vote in the Commission it would be open to a dissenting whaling country to object, and so not be bound by the decision, or even to resign from the Convention.

The Norwegians and British supported a reduction of the overall catch ceiling from 15,000 blue whale units to 4,000, in line with the Committee of Three's advice. This meant a large reduction in the Norwegian fleets, but for us it was extinction. *Harvester* could probably survive on 500 units, but not on 200, 5% of 4,000. We made our government fully aware of that, but we could not press it to abandon a policy that it had been following with our full support since 1945. Harold had preached conservation since the 1930's, balancing it skilfully with the defence of Salvesen's interests. Now the claims of conservation were overwhelming and there was almost no British interest left to defend. Extinction was going to come pretty soon anyway.

The British delegate was now Hugh Gardner, the new Fisheries Secretary, in succession to Basil Engholm, who had played a leading part in the later quota negotiations. Harold had sometimes felt that Engholm had taken too much initiative at the expense of British interests. One indignant letter to the Ministry from Harold was signed 'Yours no longer respectfully'. Gardner, who had moved from a calmer world of drainage and inland waterways, seemed an innocent in the skulduggery of Antarctic whaling, but Ian Graham stripped him of illusions and he played his part well.

There was a dry run in the Technical Committee immediately before the main session of the Commission in June 1963. Gardner courageously proposed a ceiling of 4,000 units and that was supported by Sjaastad, still leading the Norwegian delegation. The Japanese countered with 10,000, the Russians with 12,000, and the Dutch with 9,000. As no agreement was possible the debate was deferred to the plenary session.

In their arguments for the higher levels the three countries all stressed the problems for their industries if the catch ceiling was suddenly and drastically reduced. That was honest enough. But the Russians and Japanese also cast doubt on the Three Wise Men. The Japanese started by proposing a level of 10,000 units for three years, to be afterwards reviewed 'in the light of the scientific evidence available'. The Russians suggested that bad weather and accidents had pulled down the 1962/63 catch and that the true stock position was much better than it looked.

The claims, whether put forward honestly or dishonestly, that the Commitee of Three's findings could not be accepted without more evidence, and that the 1962/63 catch had proved them wrong, seemed to us, on the other side, highly discreditable to both those countries.

When the issue came before the Commission it was clear that the USSR and Japan would not accept less than 10,000 units. Even if the 4,000-unit level proposed had been put up and passed by a majority, those two countries, and the Netherlands, would have objected and the catch ceiling would have remained at the previous season's level of 15,000. Rather than let that happen the UK delegate switched to supporting the motion for 10,000 and it was carried. That was the only realistic outcome, though it spelt final disaster for the whale stocks and so for the whole whaling industry.

The agreement for 10,000 units left us still in the game with 500 units overall for *Harvester* for 1963/64, though it was doubtful whether we could actually catch our full quota. Beyond that season we could not see the expedition scraping a living from the few surviving whales, regardless of the central régime. From the moment of publication of the Committee report we had been thinking urgently about the future. *Venturer* had been successfully sold to the Japanese in summer 1962. That could not be repeated for *Harvester*, since the Japanese companies were no longer ready to pay large sums for a factory ship and its quota. We thought of throwing in our lot with the Norwegians and operating in a joint framework, but our single quota was not attractive enough to them to justify reorganising their industry on a common basis. The national quota agreement provided that factory ships might only be transferred if a proportionate share of the seller's quota was attached to it, and at this stage, when quotas were shrinking, a buyer would only be interested in the quota, not the ship. We came to the view that, even with the short time left for whaling and for the quota agreement, a buyer would still find it worth while to pay something for the quota alone. So we put it to the Japanese that they should buy our quota rights for the next three years. *Harvester* would be sold with them, but on the expiry of the quota agreement we would buy her back at the same price. That arrangement would meet the Japanese government's objections to the permanent purchase of further factory ships, and pass to the Japanese companies an extra quota which could give their existing expeditions a valuable increase of catch. I launched the idea as we gathered in London for the IWC meeting and we negotiated as the meeting proceeded. Harold and Max had already given their blessing in Leith. We all realised that we were near the end of our great whaling venture. It was better to make a clean break now when there was still a little value in the business than to struggle on for one or two more seasons to a miserable demise. The financial results of our whaling from 1958 onwards had been poor. We had done little better than break even on average in the years up to 1962 before charging any depreciation, and we were continuing to spend capital on the new meat extract plants and second-hand catchers to strengthen the fleet.

Harvester was sold with her quota to the Japanese company Nippon Suisan Kaisha for three years for a payment of £150,000 a year, the ship to be returned to our ownership at the end of the period when the national quota agreement terminated. That was more than we could have hoped to earn for the expedition if she had gone on operating. The £450,000 was a good deal less than the £2.15m. brought in by the sale of *Venturer* a year before, even when the scrap value of *Harvester* – up to £300,000 – was added. That reflected the drastically reduced expectations for whaling after the Committee of Three report.

The provision in the quota agreement that a factory ship could only be transferred between countries with a share of the seller's quota attached to it was intended to prevent quotaless expeditions coming into the hands of signatory countries, or, worse, into the possession of other countries which might operate them outside the agreement and destroy it. There were some initial doubts expressed from Norway as to whether this was a genuine sale, since the ship was to be passed back to Salvesen after three years. In terms of the agreement however it was a proper transfer of whaling material and quota and what happened after the expiry of the three years was irrelevant.

The implication of our contract with the Japanese company was that at the end of the agreement we would again become owners of *Harvester* together with any claims that the original British right to a 5% quota might give in any new agreement. It seemed highly unlikely that we would want to return to whaling then, or that there would be national or company rights of any value, but we hung on to our nine best catchers for the three years, while *Harvester* lay moored outside Tønsberg, sad and shabby, and we made sure that our government continued to defend the British interest. By 1966, with the total catch limit now reduced to 3,500 units and whale stocks almost fished out, it was clear that there was no return. *Harvester* was scrapped, the catchers broken up or sold, and the British government was able to stand down. The UK remained active in the IWC, and, to the dismay of its former friends, Norway and Japan, became a leading member in the moral movement to ban whaling completely.

Leith Harbour had been closed at the end of the 1960/61 season. In spite of the low catches of our last season it still looked attractive to Japanese whalers and in 1963/64 we leased the station to Nippon Suisan Kaisha. They were pleased enough with the results to take the station for another three years at a rate not much lower than they had paid for *Harvester*'s quota. In the event they gave up after only two years because of falling catch, but they had done well with their purchase of the *Harvester* quota so there were no regrets on either side.

Our whaling account in 1963 gave us a profit of £800,000. The last season had been quite a good one, meat extract prices had recovered, and we had the once only benefits of not having to repair the fleet and of selling off stores. For the next four years good income from the *Harvester* and Leith Harbour deals eased the pain of parting.

Apart from the passive involvement at South Georgia our whaling was now over. As soon as the *Harvester* deal was completed in July 1963 we set about dismantling our operations. We still had about forty whalecatchers, some of which had been left, moored at Stromness and Jericho, as not worth sailing home. Several were sunk by a particularly heavy fall of snow during winter 1964 and still lie there. Many of the boats we had brought home were sold and converted for fishing in the Norwegian and Danish fleets. A few continued as whalers in South Africa and Iceland. One, *Southern Actor*, fished off Spain. She was discovered years later laid up there and brought back to Sandefjord to be restored as a whaling monument. In summer 1995, fully restored and in her Salvesen colours, she was handed over to the Sandefjord whaling museum. The bell on the bridge was from *Southern Hunter*, recovered when she was wrecked at Deception in 1956 and brought back to me by Captain Bartho. On the celebration trip Ole Christensen, gunner on the boat for eleven Antarctic seasons, returned to the bow at the age of eighty-four and shot a harpoon at a symbolic whale.

There were considerable stocks and spares which had to be dispersed, although the spread in the withdrawal of the three expeditions helped to limit the surpluses left on our hands. Harpoons, ropes and wires were gladly taken up by the remaining Norwegian expeditions. More routine ships' stores found their way to our merchant navy ships. Our considerable stock of blankets went to warm destitute Moroccans. Arvid Rynning, king of the Leith Harbour winterers, took over the slop chest stocks and peddled sheath knives, Parker pens, cine cameras and shaving cream round Tønsberg for several years.

The dispersion of our whalers turned out to be less painful than we had feared. In the last full season of our three expeditions, 1960/61, we had 1,626 men in the Antarctic, 704 British and 922 Norwegian. In *Harvester*'s last season the figures were 574: 259 British and 315 Norwegian. The run-down of the expeditions over three years had made it easier for our redundant whalers to find jobs. Some of them joined Norwegian whaling companies for their few remaining seasons, many transferred to the merchant fleets of Norway and Scotland, and others went back to crofts and farms. Whalers, and Norwegian whalers in particular, were not fussy about what they turned to. Two of our gunners settled down happily for years as ABs on the Moss-Horten ferry in Oslofjord.

We set up a fund in Norway to give money to elderly ex-whalers where they needed it; the sums were modest but seemed to be well appreciated. In Scotland we created a trust for our Shetland whalers to make grants and loans to any of them needing help in new employment such as fishing, knitting, or crofting. The Shetland Trust was a notable success, mainly because we put it completely in the hands of James Tait and his law firm in Lerwick, leaving him to make payments at his discretion. A lot of useful activity was helped and there were no bad debts. The Shetland Trust still continues, now available to everyone in the Shetlands and combined with the vastly greater sums earned more recently from the offshore oil industry. We made a similar trust within the Highland Fund to provide for other whalers from the north of Scotland.

It would have been ideal to absorb our whalers into the various new enterprises which we were developing. They might have become operators on our floating fish factories, cold store foremen, or drivers of long-distance lorries. A few of them did make the switch successfully, but generally their background and way of doing things were so different from the practices of our new industries that they could not fit in. Only in Peru, where the fishmeal industry was just a short step from whaling, did we have a fruitful follow on.

The loyalties of whaling run deep. Thirty-three years after *Harvester* last returned from the Antarctic (carrying also our last consignment of penguins for Edinburgh Zoo) there was still a lively Salvesen Whalers' Club based in Edinburgh and a larger one in Tønsberg. In the autumn of 1995 the British contingent visited Norway and banqueted with their Norwegian colleagues on whale steak and *akvavit*.

In the final whaling dispositions the shareholders were not forgotten. The board decided that it had ample reserves for new business and made a capital distribution of £1m. Harold had retired from the management company, Christian Salvesen & Co., in 1960, when he was sixty-three. He had always been highly critical of old men in business, believing strongly that people were well past their best by that age and should retire gracefully instead of hanging on and stopping younger people from coming forward. He had acted in character, passing responsibility to Max and myself, though remaining a powerful and explosive presence behind us. He was still chairman of the South Georgia Company board, and fought hard against any pay out to shareholders. He maintained that the extra money that had come in from the winding up of whaling should be kept in the company and not dissipated in that way, but the non-executive directors, his two brothers and a cousin, who represented more directly the Salvesen family shareholders, could not accept that. Harold was outvoted for the first and only time, and, though indignantly, accepted defeat shortly afterwards. He resigned from the chair, passing it on to Max Harper Gow.

It was appropriate that Harold's retirement should coincide with the end of whaling. He had put his life into it and towards the end it sometimes seemed to us that he would not be averse to pulling down the pillars, like Samson, and dying in the ruins. The build up of our whaling fleet before the war and after, the loyalty and pride of our whalers, the complex chain of negotiations down the years between companies and governments to preserve the industry, all bear the stamp of his talents of forceful leadership, managerial ability, and sheer intellect. His obsessive search for truth and principle would have put him at home among politicians and divines of the nineteenth century. But that was linked with an unusual degree of aggressive self-assertion. He was an intensely competitive person, determined always to display his superiority, even when it was already manifest. This came out in perpetual battles with anyone who might seem to challenge his intellect or principles. There were very few people who attracted his respect, and many whom he condemned as mentally or morally inadequate. He was contemptuous of civil servants and scornful of competitors; he despised most scientists as second-rate. So the company tended to steer a lonely course, though sometimes his instinct to do the opposite to everyone else paid off. Within the company some of the new partners whom he introduced after the war found it impossible to work constructively with him and either left or were crushed into passivity. Max and I, a new generation, had developed mechanisms of defence by the time we came to take responsibility, but our day-to-day management was orchestrated with perpetual arguments, sometimes about high policy, sometimes on trivial matters of administration, which took up a great deal of nervous energy. They never damaged the company: Harold, having triumphantly come out on top of the argument, often modified his view and took as his own the policy that he had been passionately arguing against. In his self-critical moments he used to quote to me the philosopher David Hume that 'Reason is, and ought to be, the slave of passion'. Most readers of Hume take that to mean that passions or feelings set your goals and reason tells you how to get there. Harold's interpretation was that reason provides you with the arguments to justify doing what you have independently decided is right.

Harold had strong views about almost everything, and he rarely concealed them. In the public domain they were broad and liberal, though he was unhappy if they appeared to coincide with accepted wisdom. His special loyalties were expressed in the large gifts he made from his fortune to universities and for the welfare of merchant seamen. Such generosity was the outer expression of warm sympathies which guided much of his life, and without which he could not have enjoyed the love and respect of his family and friends, including of course myself. But he did not care to parade these qualities when intellectual battles were to be won or the company defended. He quoted to me with pride the appeal made to him by Ingvald Haugen, leader of the Norwegian whaling unions, during a particularly difficult negotiation. 'Vaer human, Salvesen' – 'Be human'. (I refrained from pointing out that he had mistranslated. Haugen had asked him to be humane, not human.)

The Salvesen family, after its migration to Scotland, kept strong family ties with kinsmen in Mandal. In 1930 a family book came out recording the descendants of Christian's great-grandfather, Salve Aanonsen, with their biographies. Harold inherited this interest and searched for further material on his ancestors, to prove the original editor, Julius Salvesen, inaccurate or careless in his work. Harold was also, predictably, steeped in the company's history. He did a good deal of research himself into Salvesen's activities in the nineteenth century, which included starting a debate on whether the ship *Olav Kyrre*, commissioned for the Mandal Paraffinfabrikk in 1867, could be claimed as the world's first tanker. (It couldn't.) He put the company papers and his own work safely into the Edinburgh University archives, making sure that when the company history came to be written the necessary material would be to hand. He had seriously intended to write the history himself after his retirement, but he found the task of composing to his own standards of accuracy impossible. Two years of work produced half a dozen pages of tortuous prose, loaded with round and square brackets, reservations and footnotes. He wisely handed over to a professional historian, Wray Vamplew. He died in 1970, when the historian had barely begun his work, and it fell to me to oversee its completion. *Salvesen of Leith* appeared in 1975, an account of the activities of the company in shipping and whaling and other enterprises up to 1945. Our whaling was already ten years in the past when the book was written, but we decided that it was too soon to write up the later years. Harold would certainly have considered it a very slipshod work, but we did our best. Fortunately the magisterial Norwegian *History of Modern Whaling* was well under way before he died. The author, Johannes Tønnesen, was discovered to be a distant relative from South Norway. Harold was able to pass on to him a great deal of material about the company from both the remote and the recent past. So Salvesen figures handsomely in a history which might otherwise have given too Norwegian-centred a view.

Our withdrawal from whaling proved to be well timed. For the 1963/64 season the IWC had set a ceiling of 10,000 blue whale units, but the actual catch fell short of that by 15%, showing that the growing scarcity of whales was imposing its own catch limitation. In the following years the Commission reduced the catch ceiling each season, though never enough to match the falling stock assessments of the scientists who were continuing the work of the Three Wise Men. By 1970 the catch ceiling had been brought right down to 2,700 units, still probably more than the sustainable yield from the severely depleted stocks of fin and sei whales. Meanwhile the Norwegians and Dutch had given up, and the Japanese and Russians had reduced to three expeditions each. The last Norwegian expedition was laid up after the 1967/68 season; it was *Kosmos IV*, previously *Walter Rau*, built in 1936 in the German drive to enter Antarctic whaling. Eventually, in 1976 and 1978, catch of fin and sei whales was totally banned by the Commission. The Russians and Japanese were left to do what they could with minke whales, a much smaller species, previously ignored by whalers. Minkes were abundant, the population having evidently increased to fill the gap left by the larger species.

The misreporting of its catches by the USSR continued to bedevil the efforts of the IWC and its attendant scientists to keep their stock assessments up to date. The real figures for their catches were released, though not officially published, in 1992. These confirmed that the USSR had ignored the Convention; their expeditions had regularly fished well beyond the agreed international and national quotas, and they had continued to take large numbers of blues and humpbacks after their catching had been totally banned. So blue whales, which it had been hoped would recover after being fully protected, continued to decline and the measures to protect fin and sei were frustrated by the continued high catch. With the removal of the European expeditions the Soviet catch became a substantial part of the total, so the impact of false figures was considerable. This situation only ended in 1972, when the International Observer Agreement was at last put into operation.

The final run down of the Norwegian and Dutch whaling fleets was similar to our own but a little later. Their best floating factories had an extension of life in marine offshore activities. *Barendsz* and *Kosmos V* (Jahre's intended third floating factory which was denied permission to operate by the Norwegian government) became fishmeal factories operating successfully on sardines off south-west Africa. *Thorshavet* worked similarly on horse mackerel off the West African coast. *Thorshøvdi* became an oil drill ship, and *Juan Peron*, another frustrated floating factory, was cut in two horizontally to make two large drilling barges. We would have been pleased to get *Harvester* into one of those activities after her return from Japanese ownership, but her steam engines and smaller size made her less attractive.

During the following years the Commission changed its interest from the regulated catching of whales to the broader issue of whether whales should be caught at all. It attracted the attention of preservation bodies such as the World Wildlife Fund, and its membership was increased by countries which had no interest in whaling but felt they had a world duty to protect whales as an endangered species. The whale, as the largest of the earth's mammals, has always had a mystic attraction, which has latterly found its expression in a growing and widespread view that it is wrong to hunt and kill them. In 1998 the controversy is still going on, with the Japanese and Norwegians still intent on harvesting minke and other small whales off their coasts, while most other nations are trying to stop them.

We should not write off the International Whaling Commission as a total failure. It failed to conserve the whale stocks, but without it the depletion would have been much more drastic, and some stocks might have been wiped out completely. The rules it introduced for minimum length of whales, length of season, initial protection of individual species and overall catch ceiling all helped to reduce the impact of intensive catching, and that régime covered not only the Antarctic but whaling fields all over the world. The weakness of the Convention, common to all international treaties between nation states, was that it could only work if there was universal and honest determination behind it. Those qualities were notoriously absent. Its history warns us that if we want to conserve natural resources, long-term benefits need to weigh heavier than short-term interests, and national claims have to give ground to international goals. In the conservation of the other great source of wealth in the sea, fish, we are still a long way away from those obvious principles.

There was an epilogue to our history at South Georgia. We were reluctant to abandon the stations there completely since they could give an ideal base for fishing, mineral or defence activities. We had already looked at the fishing round the island: there was plenty of notothenia, which we caught in the fjords to supply the station, but not enough to support a freezing station or fish meal plant on shore. During the 1970s there was some fishing activity, when Russian, Polish and East German factory ships came down and sometimes made good catches. We heard that the factory ships called in occasionally at Leith and Stromness to take on fresh water, while their crews also helped themselves to anything moveable, as well as smashing most of the windows. But, as we had been warned by the fish population experts at FAO, there was not enough for a sustainable fishery. There were no minerals on the island and, unlike the Falklands, there was no potential for offshore oil. The UK defence needs in the area were covered by a single frigate and no large naval base was required.

With no sensible use of the stations in sight we eventually decided to dismantle them for scrap. But first we approached Argentina de Pesca, now called Albion Star, to see if they would sell us their stations at Grytviken and Husvik, also now abandoned. They were happy to get rid of them. At about the same time we found a company in Buenos Aires interested in buying the stations; it proposed to salvage some of the machinery and the floating docks, removing the rest for sale as scrap steel. We decided that we would scrap Leith, Husvik and Stromness, but leave Grytviken intact in case some activity appeared in the future. Grytviken was also less ravaged than the other stations, being under the eye of the government colony across the bay.

We brought this idea to the Foreign Office in 1978. The station sites had from the beginning been leased from government on a yearly basis, but the buildings and machinery were our property and we were free to remove and sell them. The government was sensitive, as it had been for many years, to Argentine claims for the Falklands and South Georgia, and we did not want our minor tidying operation to cause trouble. The officials did not welcome the idea. They said the Falklanders would not like Argentinians carrying out operations at South Georgia. I pointed out that Argentina was already responsible for fuel supplies and air services to the Falklands and I did not see why South Georgia, in which the Falklanders had never shown the slightest interest, should be treated any differently. I also argued that part of the package was the transfer of two stations on the island from Argentine to British ownership, which they must welcome. After further consideration they had no objection to our going ahead as planned. So in 1980 we signed an agreement with Sr Constantino Davidoff to sell the stations at Leith, Stromness and Husvik for £115,000. The first contract, in the form of an option, was made in September 1979, and full agreement followed in August 1980. The agreement gave him till 31 March 1983 to remove what he wanted. The only exemption from Davidoff's rights to the stations, inserted by me on archaeological grounds, was the old whalecatcher *Karakatta*, still sitting on the beach at Husvik. I thought it might be appreciated by future generations as a historic relic.

The Falklands' governor was naturally anxious to keep the operation under supervision, so Davidoff was asked to keep him advised on his programme and was reminded of the laws of the colony, mainly on the formal clearing in of ships and the protection of wildlife.

His enterprise was slow to start, as he could not find a suitable ship to carry his men down to start dismantling. Eventually in March 1982 a party from Argentina was sent down on an Argentinian naval transport, *Bahia Buen Suceso*. So far everything had gone as anticipated, but the position was suddenly transformed when *Buen Suceso* arrived at Leith without giving any of the notice required to the magistrate at Grytviken. Its crew came ashore with the demolition workers and hoisted the Argentinian flag over the station. Thus started the Falklands War. *Buen Suceso* left Leith but was immediately replaced by the much larger *Bahia Paraiso*, which landed troops and equipment. On 2 April the Argentine forces invaded the Falklands.

It seemed that Davidoff was an innocent party in this. In seeking naval transport for his workers he provided an ideal pretext for an incident at South Georgia to raise the curtain for the Falklands invasion already planned. Much was made in the British press about the huge sums of money Davidoff stood to gain from his enterprise, '£7m. before expenses', was stated in the *Sunday Times* book *The Falklands War*. In fact the yield from the stations would have been quite modest. Even with the low costs of an operation from Argentina Davidoff would have done well to clear £200,000 on the deal. After the war ended in June, Davidoff astonished us by telexing to ask when he could go back to South Georgia and start work, as if the earth-shaking events of the past three months had never happened. Commercially he was within his rights, as the demolition contract still held, but he seemed not to realise that no Argentinian was going to be allowed to get within sight of the island for a long time. The frustration of the contract was of his own making. We let it lapse, and kept the money he had paid us two years before.

We still wanted dismantling work at the stations to be done and tried to find interest from elsewhere, but without success. Meanwhile some of the scientists of the British Antarctic Survey, which had a base at King Edward Cove, started to report small leaks from the fuel oil tanks on the stations. Most of the oil had been removed at the end of whaling, but a little had been left behind. The governor of the Falklands was worried about this and called on us to take action. The Antarctic with its unique bird and fish life must be kept unpolluted, particularly now that Greenpeace was interested in the area and Argentina would exploit anything that looked discreditable to Britain. I took a hard-bitten whaler's line. The leakage was tiny, a little viscous fuel oil oozing out on to the ground round the tanks. Even if all the remaining oil in the tanks poured out into the sea it would quickly disperse with negligible effect. The leaking places were pipe joints and, in one case, bullet holes from the Argentine invasion. I thought two men could deal with it with plugs and cement boxes in a couple of days. Why not put the army garrison at Grytviken on to it? The Foreign Office didn't like that reply. Soon afterwards I retired, leaving the argument to my successor. He, taking a more detached and perhaps longer-term view, decided not to fight the government. An expedition, jointly funded by the government and ourselves, went down to remove the oil, which had to be heated and pumped away. It also tidied the stations to what were considered environmental standards, including the burying of asbestos sheets, no doubt considered a health hazard for sea-elephants. With that done we were able to hand over the stations to government and divest ourselves of further responsibility.

Epilogue

Our company went on from whaling into many new activities, most of them away from the sea. The Norwegian whaling companies invested in merchant shipping, particularly tankers, a very successful move, though not destined to give them a lasting base. We saw shipping as a declining industry in Britain and were not attracted to great expansion there, though we kept up our merchant fleet and only left the sea for good in 1990. One road we took was into the development of fish-freezing factory ships, which we started as a diversification while whaling was still at its height. The history of that pioneering enterprise is vividly told in Jack Campbell's *The Fairtry Story*. Deep-sea fishing brought us into cold storage and refrigerated transport ashore, which became the basis of much of our subsequent activities. We also moved successfully into industrial fishing for fishmeal, finding that in Peru the skills of whaling could be fruitfully applied to the anchoveta of the Humboldt current, but that there, as on other coasts with fish abundance, the same conservation problems arose and were no better resolved. In the thirty years that have pased since the end of our whaling Salvesen has grown steadily and continues to flourish as a public company, though now in areas far removed from our original trades.

Companies have to adapt their activities and their characters in each generation, and the rate of change may now be even quicker than that. The whaling that made Salvesen unique in the first half of the century had to go, not because the industry was obsolete, but because the stock of whales ran down and none of us were clever or far-sighted enough to devise ways of maintaining them as a steady source of wealth.

Salvesen's history in whaling and fishing must be considered in the context of its period, when the globe was still to be fully explored and the discovery of new natural resources on land and sea remained an important mainspring for the creation of wealth and welfare. The romance of adventure masked the physical and financial risks for those who undertook them.

We live now in a different era. We realise that human activities on the globe have developed to a level at which their impact on our environment can seriously affect the welfare, and even the survival, of the human race. We have become almost as concerned about animal lives as about our own, and in particular those animals which breed outside our domestic control. Some of the new principles rest on strong practical arguments, others, like the unqualified preservation of whales, have become moral issues, no less powerful for that. We shall find the glory of enterprise in coming years not in far-flung expeditions to uncharted seas but in the imaginative conversion to fruitful purposes of the work of the software designer, microbiologist or physicist.

Appendix

BOOKS ON SALVESEN'S ANTARCTIC WHALING MENTIONED IN THE TEXT

Den Moderne Hvalfangsts Historie, vols. 1–4, Norges Hvalfangstforbund, 1959-70 (Arne Odd Johnsen, vol. 1; Joh. N. Tønnessen, vols. 2–4). This contains, at the end of vol. 4, a full bibliography of the copious literature on Antarctic whaling.

The History of Modern Whaling, C. Hurst, 1982 Abridged translation of the above in one volume. Also contains bibliography.

Wray Vamplew, *Salvesen of Leith*, Scottish Academic Press, 1975. The official company history, covering mainly 1846–1945.

Pa Hvalfangst med Salvesen, ed. Tore Asplin, Tønsberg 1974. Taped interviews with Norwegian and British whalers on their whaling experience with Salvesen.

The Antarctic Region

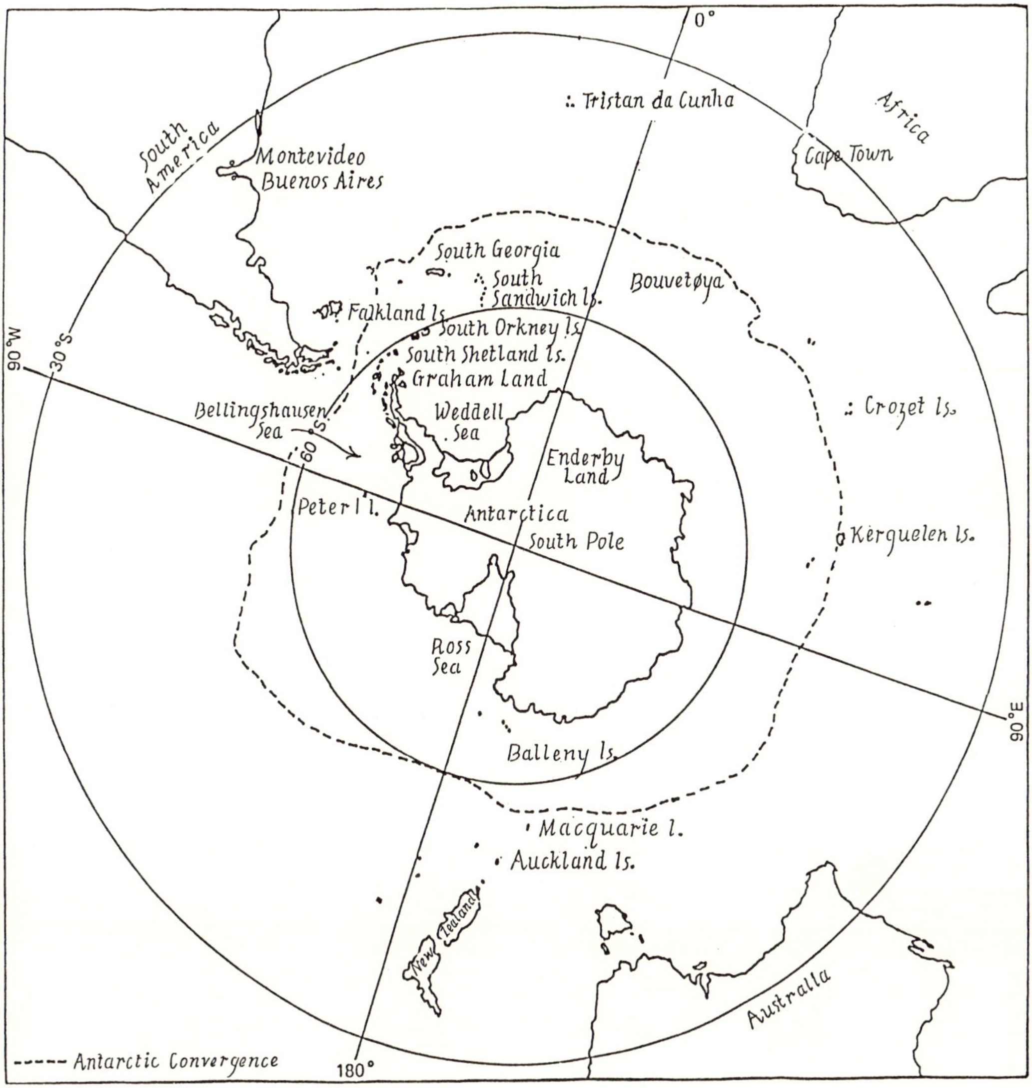
0°
Tristan da Cunha
Africa
Cape Town
South America
Montevideo
Buenos Aires
South Georgia
South Sandwich Is.
Bouvetøya
Falkland Is.
South Orkney Is.
South Shetland Is.
Graham Land
90°W
30°S
Crozet Is.
Bellingshausen Sea
Weddell Sea
60°S
Enderby Land
Peter I I.
Antarctica
South Pole
Kerguelen Is.
Ross Sea
90°E
Balleny Is.
Macquarie I.
Auckland Is.
New Zealand
Australia
----- Antarctic Convergence
180°

Index